The New Macro-Cosmic World of Order

Merlina Marcan

ISBN 978-0-646-80692-1

Disclaimer
Whilst every effort has been made to ensure that the information contained in this book was correct at the time of publication, no liability will be accepted by the author for the accuracy or completeness of the information. To the best of the author's knowledge, the material used is free of copyright, or has been acknowledged. An apology is extended for the use of any material unwittingly unacknowledged.

Prepared for publication by

www.minipublishing.com.au

SIMPLY…
THIS BOOK IS DEDICATED TO ALL THE
BELIEVERS OF 'THE ETERNAL COSMOS'
AND TO ALL THOSE WHO SOON WILL BE.

IF YOU ARE NOT ONE TO FOLLOW YOUR
HEART,
THEN SIMPLY FOLLOW THE FACTS WITH
AN OPEN MIND,
AND MAYBE THEN…
YOUR HEART WILL FOLLOW.
…SEEK AND YOU WILL FIND THAT…

"WE ARE NOT ALONE"

Merlina Marcan

CONTENTS

Part 1:	Preface……	6
	The Author……	10
Part 2:	Introduction……	12
Part 3:	Earth…our slice of Paradise……	15
Part 4:	Orbs and Digital Cameras……	18
Part 5:	My Journey to Discovery……	21
	What are Cherubim?……	24
	Archangel Michael……	26
	Author's note:……	38
Part 6:	Orbs…they select & choose 'Where and When' to appear	41
Part 7:	Colours, Size, Shapes & Direction of orbs……	62
Part 8:	Categorization of Orbs……	75
Part 9:	Pareidolia & Matrixing……	82
Part 10:	Transcendent Module 1 orbs (TrM1 orbs)……	87
Part 11:	Transcendent Module 2 orbs (TrM2 orbs)……	96
Part 12:	Transcendent Module 3 orbs (TrM3 orbs)……	108
Part 13:	Where are they coming from?……	112
Part 14:	Case Studies:……	121
	Case study 1 (Diana Davatgar)……	124
	Case study 2 (Peter F Kahrs)……	128
	Case study 3 (Cathy Finch)……	133
	Case study 4 (Foco Tonal-Hunab Ku)……	138
	Case study 5 (Theresa Kaplan Amuso)……	143
	Case study 6 (Marion Atehsa Cyrus)……	148
	Case study 7 (Andrea Corsick)……	153
Part 15:	Metaphysics vs. Physics vs. Scientific Research……	161
Part 16:	My Research……	172
Part 17:	Real image of Sentient…'and my paintings'……	202
Part 18:	Conclusion……	205
	Orb photo credits:……	207

Part 1
PREFACE
My Opening Chapter

It has taken a lot of thought into finding the appropriate opening chapter for this book, and also to find the appropriate '*title name*' for this book.
How does someone begin to talk about a particular topic (as in this case the Orb Phenomena), that has kept everyone baffled for decades, as to what they could be, and how, where or why, they are appearing?
The Orb Phenomena is a topic, which obviously does not have much accessible information, or guidelines about it, there are no textbooks written about them that can be leant in schools, colleges or universities. There are no manuals or instructions to read about these phenomena that may have some proven universal knowledge or concrete evidence about them. And so far, most of the information that has been written about them by some pioneering orb researchers is considered actually anecdotal in nature by mainstream science.
But in effect, in some way, this topic in the near future, could perhaps change the course of history, and change the course of human reality.
So…I will ask this question again!
" HOW does someone begin to talk about an unpredictable, uncertain, unknown, controversial and mystical topic such as Orbs"??
(I searched on the Internet to find out more about '*What Criteria is needed for true Scientific Research?*')
How may you ask??... I will tell you how!

With solid photographic evidence that can be replicated over and over again, anywhere and anytime, even under the most effective investigating procedures of controlled experimentation and data analysis.
That's how!

If an object cannot be seen with the naked eye, but is believed to exist in a certain state outside of our own normal comprehension, then IT IS A SCIENTIFIC NECESSITY TO FIND A WAY TO PHOTOGRAPH IT!
Evidence through vision and observation is a very powerful research tool in science.

So therefore, I have discovered a photographic procedure that I have named **'MFT Method', (**which are the acronyms for Merlina Focus Technique). This method is a process, which engages a simple procedure that can focus into these multi-dimensional Electro Magnetic Fields that we see externally as 'orbs'.

The burden of proof lies on the one who makes the claim, and until this day, no one has come up with plausible visual proof or solid groundbreaking evidence that '*genuine orbs are carriers of much detailed data and information inside them*'.

I am not a scientist, I am a *Citizen Researcher,* and this book is about my own personal research, which engages in my pioneering photographic procedure (MFT Method) that has never been ventured before. And of course, all my hypotheses are based on genuine and authentic images that I have captured *inside these energy fields.*

My hypotheses may be challenging to many people, (especially to main-stream science), but they are based solely and entirely on exclusive photos of my data files over a nine year period of my research.
My research work is directed in a specified broad area of the 'Orb Phenomena' and will provide a broad base of knowledge through 'visual evidence'. It will acquire new knowledge and new findings with specific results, predictions and conclusions.
My discovery is a simple procedure that cannot be dismissed as fake!
I believe that when my research is tried and tested, the controversy around the 'Orb Phenomena' will be proven to be one of the most extraordinary and significant phenomena (outside of our human reality), that mankind has ever witnessed the likes of.

As for the *'title-name of my book',* let us consider a flashback and go back in time into the **'early 17th Century'**. In this era of time, no one could conceive or conceptualize the slightest possibility for the existence of *a living dimension so tiny and minute that was invisible to the naked eye*. In this era of time, no one could surmise or envision an un-seeable minute microbial dimension that could sometimes be a threat to mankind, but also on the flip side of the coin, it could also be a blessing by finding cures, and wiping out epidemics and many diseases in humanity. The thought of such a *'ridiculous idea'* of the existence of a 'microscopic dimension' was laughed at, it was scorned at, and it was ridiculed by the scholars, scientists and mainstream science of that time.

But a common man by the name of *'Anton van Leeuwenhoek' (who was not a scientist)* was to become the first man to observe bacteria and protozoa by using a simple single lens. Leeuwenhoek was not an educated man, and his studies lacked the organization of formal scientific research, but his passion for the unknown, and his power of 'careful observations' enabled him to make many discoveries of enormous fundamental importance that was to change the scientific history of mankind. His *'observations'* helped lay the foundations for science in bacteriology, microbiology, protozoology, and medicine in general. With his new discoveries, he managed to change all the existing beliefs of that era, and opened new doorways in science by discovering the pioneering *'Microscopic World'.*

*The '**MICRO-SCOPIC World**' became a technical term used to view tiny organisms that could not be seen with naked eye, and this discovery was to change the scientific dogma of that era, it opened up new unexplored doorways to science, and medical breakthroughs, and gave a new hope of survival to mankind.*
When we 'focus' into these micro-dimensions, we can see and view another world of great visual detail and information. So having said this, just like the microscopic dimensions, let us now consider another mysterious and mystical world with great visual detail and information we cannot see with the naked eye…
I have named this world, the new 'MACRO-COSMIC World'.

This new macrocosmic world is overflowing in our universe, our galaxies and our star systems. It is enriched with so much information and so much detail, which can be found in the Orb phenomena as they travel, glide and slide effortlessly through our cosmos.
Hence, the title of my book: ***"The NEW MACRO-COSMIC WORLD OF ORDER".***

I will finish my opening chapter with two amazing quotes that deeply resonate inside my understandings of Knowledge for Truth:

QUOTE 1:

"TRUTH PASSES THROUGH THREE STAGES BEFORE IT IS RECOGNISED",
First, it is ridiculed
Second, it is violently opposed, and
Third, it is accepted as being self-evident".

Arthur Schopenhauer

QUOTE 2:

"Sometimes one person can make a difference in the world.
In fact, it is always because of one person that all the changes that matter in the world come about. So be that one person".

"YOU NEVER CHANGE THINGS BY FIGHTING THE EXISTING REALITY.
TO CHANGE SOMETHING, BUILD A NEW MODEL THAT MAKES THE EXISTING MODEL OBSOLETE".

Buckminster Fuller

THE AUTHOR

Merlina Marcan was born in Athens (Greece) and at the age of only six months old, her parents emigrated to Australia where they have lived in the City of Sydney ever since.
Growing up as a normal teenager in the Greek-Australian community, Merlina finished her Higher School Certificate and she also finished her studies of Pathology and Microbiology at the Institute of Technology in Sydney. After her initial training at the Sydney General Hospital in the late 1970's, she went to Europe to do further training in hospitals and private clinics. A few years later, Merlina decided to abandon her work in the medical field and move onto new ventures in the entertainment world by following her heart and her free-spirited personality, she took on a new career as a singer and entertainer.
In the mod-world of the 1980's and 1990's she travelled around the globe as an entertainer with her musical band and choreography dancers to countries such as Greece, Australia, Germany, Egypt and Canada, only to mention a few.

But…In 2007, due to an intense and emotionally immense family crisis, she began to question her priorities and importance of life. She began to undergo her transformation from a 'skeptic-agnostic' to becoming aware of her 'spiritual dimension'. She began to seek answers to existential and spiritual questions and trying to connect with her guardian angels and her spirit guides for help in this critical time of her life.
In 2009, Merlina was still dealing with her family trauma and she decided to end her singing career. She began a hobby of amateur photography and spent nearly every night photographing orbs. Capturing orbs with her camera became a fascination to Merlina; she felt that it had a soothing effect on her, and she always had a camera with her, photographing them every chance she could get.
In June of 2011, Merlina was to experience a moment of *enlightened spiritual awareness* that was to change her inner-self and her belief system. After a series of serendipity events that were to change her life, with the help and guidance of what she believes is from a *Higher Divine* to help her in her mission, she began to better understand orbs and began to classify orbs into categories and groups, this was something that had never been done before.

Today Merlina is dedicated to the research and understanding of the 'Orb Phenomena' and is regarded as a world expert. She is the founder of the 'MFT Method', which is a cutting-edge breakthrough method of bringing forth new '*visual evidence of Sentient Beings from Other Dimensions*'. This is considered to be compelling visual evidence that may confirm that visitation to Earth is occurring by beings from other dimensions and other realities.

Merlina is a thinker, a visionary, a pioneering explorer (of undiscovered realms and undiscovered invisible life forms) and seeker of truth. This research has led Merlina to incredible discoveries, which include thousands of amazing photos of cosmic beings, from outer-space civilizations, Extraterrestrial and Spiritual Dimensions.
For close to a decade, Merlina has pioneered in the Orb Phenomena, what others would call her work undoubtedly controversial. Even though she is not a scientist, she is a Citizen Researcher, and her work is now growing public interest and awareness with her amazing visual discoveries and pioneering hypotheses to give new evidence that…
WE ARE NOT ALONE!

Part 2
Introduction

Many people have *speculated* on the possibilities for life in other worlds, other dimensions, and other frequencies, but without proof or concrete evidence in tests that can be replicated, or duplicated over and over again, this hypothesis can be easily pushed aside or debunked. Especially when such hypotheses are connected to many fears, dis-beliefs and anxieties of the unknown. It is unfortunate that some people cannot grasp, or deal with a new concept of perception, as they feel that in doing so, it would take them out of their 'comfort-zones'.
But in reality…we can create reality, by simply changing our own frequency, and if we allow our minds to open up to the cosmos and the universe, we will then allow ourselves to see and feel things that were not possible before.

There is a connection between *Soul, Spirit and Consciousness*. When we begin to operate at a higher frequency, such as in meditation or self-healing sessions for example, we shift our energy to *positive energy*, and we can radiate this energy, and shift our *aura* to a more spiritual awakening. In past history, 'auras' had generally been described as a halo, a spiritual radiance or glow that could be seen around certain special individuals. But in modern day, the aura is recognized as an electro magnetic field that surrounds 'ALL' living creatures.

With the pioneering discovery of *Kirlian Photography*, and the invent of Aura Photo Cameras, we now have the ability to measure the frequencies of resonance in a person's auric field, by converting their electromagnetic energy into equivalent frequencies of colour from the colour spectrum, which can be interpreted to indicate a person's current physical, emotional and spiritual well being. As we grow, as we go through life changes and as we develop, '*so do our auras'*. So this is why we should not be surprised when we hear suggestions about having the ability to create our own reality, because it has been scientifically proven that we really do have this ability.
Quantum Physics has confirmed that '*we are energy*', and that we can access a form of enlightenment on a somatic level by using our higher self to control and rectify any disharmonies that everyday life may bring us.
Even more intriguing may be that 'ORBS' have these extraordinary powers, as they can sometimes transmit and transfer their conscious light energies to us. And often orbs come here to hover around the meditator or the healer to help them regain a positive balance, as we will see later on.

Studies have shown that '*energy has a consciousness*', and we also now know that energy is not contained simply only in our *Earthly 3 spatial dimensions.* String theorists now believe that extra dimensions do indeed exist, and the superstring theory supports that the equations that describe this theory, requires a universe with no fewer than 10 dimensions. But for us humans, we are 3-dimensional beings, it is perhaps next to impossible to try and imagine even one single additional spatial dimension, because we do not know, and we have no idea, what a 4th, 5th or 6th spatial dimension could be like…it just boggles the mind!

We only have the capability to witness and experience only three independent co-ordinates, which are *height, width and depth*, and this is all that we can physically see. Anything beyond this to us is unknown territory. And to be more specific, if for example *a sentient being from a Higher Dimension* did appear in our 3-D world, chances are that we would most likely misinterpret the sentient beings true form, colour and shape, because **we can *not* see things as '*they are*', we can only see things through the eyes of who '*we are*'.**

So how can we ever perceive the dimensions above us? Perhaps to answer this enigma, it would be wise to say a partially true answer could lie between the realms of our consciousness, and our energetic aura shifts of our spiritual awakenings, in order to be able to understand these other dimensions a little better. And in saying this, it brings me to another question that is one of the greatest enigmas of humanity.
"***What is beyond death***"**?** Today, there are hundreds of notable world-renowned scientists that are studying the mysteries of consciousness and 'non-material' science, such as near death experiences, out of body experiences and past-life regressions, only to mention a few of the many non-material topics of research.

What is *beyond death* is a question that occupies our thoughts many times during our lifetime, because we all have to leave this world eventually. Where do we go? If our soul leaves the physical body, does it then return to the Spirit World? What happens to us?
The Spiritual World is a *'multi-dimensional structure'* that we cannot conceive or understand in our 3-D physical existence, because our physical body is merely a vehicle for the soul, and our soul is energy. Beyond death, our soul can eventually enlighten to higher dimensions…but for now we cannot see or conceptualize this because we are in our bodies and we are alive in 3-D form. This is where we hit a brick wall! Because even though the vast majority of the population in this world believes in the afterlife, (as I too very much do), it has yet to be proven with solid ground-breaking evidence that will be accepted as undeniable plausible evidence, and this is why it still remains an enigma.
Views about life and death are many and varied, and for centuries some religious beliefs have taught the doctrine of the immortality of the human soul where the onset of death opens a door to existence into another world, another reality, another dimension.

Note:
(*Orbs and the afterlife*: I believe 'orbs' are energy fields with '*conscious intentions*' from other realities that want to make their presence known to us. These conscious energy fields can *intentionally be transmitted to us* from these other realities as a form of communication. Through my research, I have come to believe that there are intentional energy fields of orbs that come from the ***Spiritual Realms***, and that there are intentional energy fields of orbs that come from the ***Alien Agenda Realms***. We will see this in great visual detail later in the book).

Part 3
Earth...our slice of Paradise

I have often asked myself, could a *Higher Consciousness* in humanity on a Universal level world-wide, create a new species, or rather, a '*new breed*' of human incarnate for a New Earth? As we now know, '*we are of energy*', and if we humans can sustain good vibrations on a universal level around our Earth, then we can all rise together to a higher platform of existence as we can start preparing for 'A New Earth'.

Consciousness has no religion. It has no borders or limitations and it has no ideology for the discrimination of race, gender, sexuality, age, nationality or skin colour.
We are one of a kind, we are one of the same race...***we are The Human Race!***
If we, the human race were made aware of our purpose in this life, and that what we 'reap' in this physical world, we will 'sow' in the forthcoming world, we would then live life to a fuller appreciation and respect, with far less ignorant, nefarious and greedy people. It is the ignorant, corrupt and greedy humans who cannot grasp the concept that penalties will be paid, and that '*Karma*' will eventually catch up with them. Our happiness or misery, for our future generations largely depends on the decisions we ourselves make.

Our beautiful *Mother Earth* is unique amongst the other planets in our Galaxy and Solar System. We are blessed to call this rare Earth our home where the conditions on this beautiful planet, are ideally suited to our species of the Human Race. It gives us the perfect ingredients for our human survival, as it is such a rich tapestry of natural resources. This slice of paradise which we call our home, is a planet of many miraculous different landscapes, different climates and seasons, which may perhaps be unlike any other planet, as it maintains an equilibrium of water, oxygen, plants, animals and is full of life. It is one of the rarest planets to have over 65% of it covered by blue oceans and water, as compared to most other planets that have next to none, hence we call our Earth the '*Blue Planet*'.

But our Mother Earth is fragile, and she is becoming more vulnerable as we are slowly killing this beautiful caring Earth that has taken care of us for many thousands of years. We are endangering her natural beauty and balance, and we are wasting this beautiful paradise we have been given as a unique gift, like no other similar gift that can be found in the entire universe.

Loss of natural resources, damage and mutilations to our eco-systems, collapsing of environmental systems, hazardous biological and hazardous chemical contaminations, greed for wealth and greed for power, will often bring about warfare and *irreversible damage* and destruction to become a monumental human tragedy. And even though we are a tiny piece of a large puzzle of the unexplored cosmos, we may be endangering not only our Earth, but also a larger area of our universal space perimeter, and put some of our '*space neighbours'* in danger too! The Universe is 'alive' with matter we know very little of. The 'fabric of space' is made up of many living energies, and it is estimated that 99% of matter in the universe is in a *plasma state*. A plasma state that is 'invisible to us', but is 'full of life'.

My research has shown me that in this large unexplored tapestry we call the 'Cosmos', *our cosmic neighbours are continuously visiting Earth*. They are highly advanced civilizations of *many a variety of different species of Space Beings*. And *some beings* that may perhaps be a lot like us in many ways, but exist in a different frequency. In this manuscript book, I will show you images of some of these 'beings'. Some may look strange and alien, some may look semi-human hybrid forms, and some may even look like they are in a human-replica form. Whatever they are, and wherever they are coming from, the fact is that they are here…*they are peacefully watching and observing us!*

These extraterrestrial civilizations may just be curious observers, or they may be gathering data about our critical times here on Earth as they may be concerned about Earth's future. They may be coming here to send us messages, to keep an eye out for us, to warn us, to protect us, and perhaps they want to be protective guardians for Earth and mankind. They may perhaps be on a mission to remind us of our purpose of living and to make us aware of what is to come if we are not careful, and to alert our human family of the great thresholds we now face. I believe that they are sending us many messages, such as in the case of '**Crop Circles**' appearing all around the world, many '**UFO sightings**' appearing all around the world, and of course the plethora of '**Orbs**' appearing everywhere around the world.

Unlike us, they are not limited by time, space, distance or size. As they are thousands of years more advanced than us, they can easily *shape-shift* and changes their atomic structures, they can partially or semi-materialize to become translucent, they can fully materialize and become opaque. And then within the blink of an eye, they can just completely de-materialize, and disappear from sight, just like that!
So whether they are Extraterrestrial, Extra-dimensional, Inter-dimensional, Spiritual Celestial Beings, or a mixture of all these, they are coming here and they are 'watching' us.

I believe that these *Highly Advanced and Intelligent Civilizations* are sending us instantaneous information by Quantum waves or Quantum fields, that are made of various undetermined light energies and other matter that is *unknown* to us humans, often using the Orb phenomena as a vehicle of transportation. My goal in this book is to show that there is so much '*intricate information*' and '*so much data'* that can be found inside these Magnetic Energy Fields that we call ORBS!
Orbs show *Physical information*, but also show *Intellectual information*, as you will see in this manuscript book.

Note:
I would like to say to the 'skeptics, the non-believers and the naysayers' who do not believe that we are being visited by space beings from other realities, that it is not for me to change or shift your perspectives of how you perceive world reality…
BUT…it is my duty *as a citizen researcher*, to show and to inform the findings of my work to those that are willingly ready to accept that *we are not alone*!
I would like to consider myself a *Universal Ambassador of Peace and Truth.* Truth is not a conspiracy, truth cannot be hidden under the carpet for *too* long, especially when our cosmic neighbours from other galaxies and other star systems are now becoming *very keen to make their presences felt. THEY WANT TO BE SEEN, and this is why we are seeing and witnessing them all around the world.*
Eventually, the truth will bring *peace to our world*, and change the world for the 'Better of Humanity'.

Part 4
Orbs and Digital Cameras

This book is to give photographic evidence that Sentient Beings from other space civilizations, and Sentient Beings from spiritual higher realms '*DO exist*', and they travel the cosmos with a strong connection to the Orb Phenomena. I believe that we have been getting visitations from them for many thousands of years, since the dawn of humanity, but only until recently over the last few decades or so, we are becoming more and more aware of them because they now '*want us to see them*'.

What separates *US* from *THEM*, or better still, what separates *THEM* from *US*, is described with two simple words…'different frequencies'. We are different frequencies or wavelengths, and thus we are in different realms, therefore we cannot see them with our naked eyes. We cannot see them with our naked eyes, because our eyes are biologically programmed in having a very limited viewing range of the Electromagnetic Spectrum. The electromagnetic spectrum, is a term used to describe the entire range of light that exists, ranging from radio waves, infra-red, ultra-violet, x-rays, gamma rays etc.

When we think of light, we mostly think of what our eyes can see! But, in reality, most of the light in the universe is in fact, *invisible* to us. We as humans just cannot see other frequencies through our eyes, and hence we cannot see *other realities.* Our human range is what we call a thin band of 'visible light' and is mostly consistent with the visible colours of the rainbow. Anything that may exist outside this narrow band of the visible spectrum, we are physically unable to see. We cannot see sentient beings that are outside our reality, as they operate in different frequencies and different light bands from us.

Most likely, the 'orb realms' may exist mainly within the '*Infrared Spectrums*', as we humans exist within the '*Visible Light Spectrum*'. So therefore we need something that can 'blend' between these two different light-bands, we need some type of optical instrument or device, which can act as a common denominator to help us see more. Hence, the *Digital Camera* is an optical device that can act as a common denominator, as it can work with the light of the visible spectrum as well as with other portions of the electromagnetic spectrum.

Cameras have evolved from many generations of photographic technology, such as calotypes, dry plates, and film cameras and to the modern day digital camera and camera phones. As camera companies evolve and revolutionize their technology, their main goals are to produce *higher definition and better image qualities*. As technology and higher definition improves, we are starting to visually witness other phenomena, and so much more than what our naked eyes can see! I do not believe that camera companies have any intentions to keep capturing 'dust related anomalies', as this would most definitely put their companies out of profits.

Sometimes '*serendipity*' happens in technology, and I would say that this is a ***Serendipity Bonus for Mankind*** with the evolving digital camera robotics, for we are now able to witness realities outside of our own, that were not easily photographable with the older previous camera technologies.

So putting it in layman's term: Once the digital camera captures the infrared image of an orb, it has the ability using its inbuilt technology to *convert* any of the low infrared light that it has captured, into images that we can see. We are actually seeing the image that has been transformed, interpreted and translated (from low infrared light to visible light) by the digital camera. So basically, new camera technology is giving us the ability to capture Orbs that have been around us all this time but were not able to be previously photographed.
So as camera technology improves...WE 'SEE' MORE!

Likewise, as the camera can automatically interpret 'the image shape' of an orb, it also automatically interprets the 'colour', or colours of the orb to us, into the various limited colours that we can see. (I can say this with certainty, because some of the sentient beings you will see in this book, will have colours that have been automatically changed and adjusted by the camera into the limited colours of the rainbow. Hence, some of the sentient beings will be 'colourfully unusual' for our familiar liking, and quite often we will see them with green, blue, red, purple and multi-coloured faces for example).

False positives and airborne anomalies in digital cameras:
Many skeptics would say that orbs appearing in digital cameras are simply just 'dust particles' or 'moisture droplets'. And yes...the fact is that sometimes images, as circular spots probably are airborne anomalies. In photography, 'backscatter' is due to the camera's flash reflecting unfocused motes of dust, or similar other airborne particles.

Dust particles are the most common airborne anomaly, together with moisture, rain droplets, humidity, snowflakes, grass pollen, spores, insects, dirty lens, even lens flare or 'sun spots' when the camera is facing directly into a very bright light source such as a light bulb or directly into the sun, and so on. And here is where sometimes people can get confused between airborne anomalies and *authentic orbs*.

I strongly recommend to anyone who is interested in Orb photography to go out and experiment with the camera under various atmospheric and environmental conditions and compare the differences. The photographer will eventually differentiate between authentic orbs and airborne contributors that can create 'false positives' in digital cameras.

Here are two examples of images from various anomalies such as dust particles and lens flare, which can often have a consistent colourful spotty, or blurry faded appearance etc.

Here is a construction site where 'dust' is all around, producing these round false images.

This is lens flare forming rainbow images, when camera is facing directly into a bright light source.

Part 5
My Journey to Discovery

During the early years of my orb photography, 'orbs' and their ability to appear in digital cameras had always fascinated me. I had spent many evenings, photographing and collecting orb images. Like many other people around the world, I too would zoom into these orbs and try to imagine what they could be, or what they could mean. At that time, I'd say that I probably had over a few hundred orbs in my album collection and I was happy enough to just accumulate them, up until the night of ***28th June 2011***. On this very special night, I captured this one very special image; an image that I had no idea that it was going to change my life.

Let me take you back to this hot summer night when I was doing what I would usually do in the evenings with my free time…yes, taking photos. On this night I was orb hunting as usual, over and around the area of my house in Athens, Greece. Unbeknownst to me at that time, this certain night was to initially ignite me to a steep up-hill journey to a new pathway of discovery.
An *unusual bright image* that I captured in my camera on this night was not round like an orb, but it was extremely luminous as it stood out with a flashing golden light effect as I took the photo, like a sparkler in the sky. I was so exited to see what I had captured, so I zoomed into this image. As I zoomed in, *an unusual emotion crept up on me,* as this sent a warm electric feeling down my spine and still does every time I think of it. I could not explain this unusual emotion that came over me, as tears, happy tear came to my eyes. So after this initial unexplained emotion and amazement of what I had captured, I decided to name this photo ***'My Angel photo'***, as my gut-instinct knew that this was 'not a little bug or little insect' next to the lens, this was something way above and beyond my knowledge!

A week later on returning back to Australia, I sat at my computer to look at all my photos that I had taken on my holiday. I'd spend night after night looking at the photos, and especially I would spend many hours just staring at 'my Angel photo' as if I was compelled to do so. Over the weeks to come, my curiosity of 'my Angel photo' grew into a stubborn commitment to push my boundaries night after night yearning to find out what it signified and what message it could be sending me. So I started to experiment and research my photos, without really knowing *how*, or *what* I had to do, or *what* I should do. It was just a matter of researching with my gut-instincts, and trying to make some sense of the results obtained, and trying to classify the accumulated results into specific patterns and characteristics. And also, trying to replicate and duplicate the data results over and over again.

I started to study them many nights till the early morning hours, this was my best time because everyone else was asleep and there were less distractions. ***Without really knowing why: I was starting to feel that even though the Orbs were visiting us in our dimension, and they could see 'us', I was somehow finding a new way to be able to see 'them' inside their dimension.***
It felt like I was being compelled to keep studying them, it felt like it was an infatuation, a passion, like an addiction, but this new addiction of mine was to trigger a new pathway to my 'MFT Method'. But of course, the discovery of my MFT Method did not happen overnight. My family and my close friends know the amount of endless hours of time and effort that I have endured in this commitment to see what is inside these mysterious structures, we call 'orbs'.

With every new step I experimented, it would take me to another and another step and so on, and I was starting to realize that orbs where 'to coin a phrase'...*like onions with many layers and many facets*, (where every layer you take away there is another layer underneath and so on). I eventually felt confident of what was needed to be done on every individual orb. But of course, at the beginning of my work, my images were extremely blurry and rather distorted with poor resolution. With time, and having examined literally many hundreds of orbs, I was gradually getting clearer and clearer images.

Eventually, I was beginning to feel that I was somehow improving and mastering my 'MFT Method' and I was feeling more confident in how and what I should do. So I went back to '*my Angel photo*' and used this method, only to have my breadth taken away with amazement of the *new images* I was now seeing:

I could see a '*human-replica profile head of a bearded, long-haired male sentient'*, behind the large body of what seemed like a 'white creature or beast with no limbs'. At first I thought this white creature might be some kind of 'Celestial Unicorn', because I could distinguish what looked like a 'horn-like protrusion that I initially thought might have been attached to the front of this white creature's head, but I was wrong, because as I continued to explore this photo further, I realized that this was not a horn-like protrusion, it was more like a '*sword or rod-type instrument'* that was perhaps being held by the long haired bearded sentient that was behind this White Beast Limbless Creature.
(See photos, PAGES 27-33)

'My Angel' was now showing me its true form, and I knew there and then, (without knowing how I knew, I just knew), that this divine majestic 'Heavenly Being' wanted to appear in my photo for a reason. Again, I was getting a warm electric hair-raising feeling down my spine like I felt when I first captured this image, and it was also around this time that I began getting funny tingling sensations on the top of my head. The tingling sensations would come and go at various times during the day and night, especially on the top right side towards the back of my head. I have been told by at least three acclaimed Psychic Mediums that these are angelic beings, perhaps my guardian angels that are stroking my hair and *whispering* into my ears. Even though I cannot directly hear them, I believe that the '***whispering into my ears***' comes out as '***thoughts in my mind***'.

For days, I would stare at this *effigy of this white-golden creature* with the head like that of a 'weird bird-horse-beast', and one night as I was staring at this photo and I was laying in my bed, I must have dozed off into sleep, because I was abruptly woken up by a loud sound in my right ear. I quickly opened my eyes, and I distinctly remember having heard *one* word. One word that was loud and clear in my ear. It was a word that I was unfamiliar with, and the word that I heard in my light sleep was '***Cherubim***'! I heard 'cherubim' *only once*, but it was abruptly loud enough to wake me up! I jumped out of bed and went straight to my computer to find out what this word meant! This one word from a voice that came to me from out of the blue in my sleep had me now wandering if this angelic vision of the photo in front of me was perhaps that of a '*cherubim*'.

What are Cherubim?

I began an intense research on 'Cherubim' and found out some of the following facts:
I will try and make these facts as simple as possible.

Firstly, *Cherubim* are the plural for *Cherub*. Over the last few centuries, there has been confusion and a mix-up between Cherubim, Cherubs and cupids, which were often depicted as small winged infant like angels in murals and paintings. One main speculation for this is that Classical and Renaissance artists had begun painting and portraying these cute chubby angels and calling them *cherubs*. This false interpretation of 'cherub' is still to this day, carried out in New Age and Modern Art.

Of course, this interpretation of cherub or cherubim can be no further than the truth. For the record, to get things into proper perspective, ***Cherub or Cherubim are colossal, huge powerful and fearsome entities or guards that rank as the Second Highest Order in the Hierarchy of the Angelic Beings.*** (Cherubim are second in line, under the *Seraphim angels* who are the highest in order under God the Creator).
It is said that the first mention of '**cherubim**' is when God placed Cherubim together with a Flaming Fiery Sword at the Garden of Eden, to be the guardians and protectors (Genesis 3:24).

So generally speaking, Cherubim are colossal hybrid formidable creatures, and according to various Scripture descriptions of them, there is some dispute about their appearance, because their physical descriptions that have been gathered from various scripture manuals *differ* from one another. These celestial 'guard dogs' as they are sometimes called, are sometimes described as a *conglomeration* of various shapes of eagles, oxen, lions and human-mixed forms. Sometimes, even the heads of mystical creatures on half-bodied winged human creatures have been portrayed as Cherubim in ancient civilizations. It is therefore certain to say, that Cherubim do not resemble that of any living creature known or seen by modern man.

In ancient history, it is certain that there were two carved cherubim above the ark and in sanctuary of *Solomon's Temple* each over 17 feet high. But, unfortunately they were lost or destroyed by the Babylonians or foreign army in 586 BC, so unfortunately, no image of them remain.

There is however, some reference of the appearance of cherubim in the '*Prophetic Vision of Ezekiel*'. The prophet Ezekiel describes them as being tetramorphic face creatures, each having four faces; of a man, a lion, an ox and an eagle covered with many wings. But again even so, Ezekiel does not say if the cherubim had the distinct body of an animal nor human. What is more, the descriptions in the Bible do not give a clear picture of what they may have looked like either.

And even though, stone carvings, sculptures, wall murals and artifacts of various similar creatures have been found in ancient Greek, Egyptian, Assyrian, Babylonian and other ancient civilizations dating back thousands of years ago portraying winged bulls like the Lamassu, and even human headed lions like the Sphinx, it is still very difficult to imagine what cherubim exactly looked like, for they are like no other living creature that we know.

Could it be possible, that 'my Angel photo', this mystic white hybrid creature-beast that I captured in my camera, be a *Celestial Cherubim*? In my mind, there is 'no doubt whatsoever', that this white celestial hybrid beast *is* one of heaven's cherubim!
But I was puzzled about the images of the '*bearded long-haired celestial being*' that could be seen '*behind*' *this white Cherub* when I experimented with my MFT Method. I was asking myself *WHOM* was this White Cherub 'escorting' through the cosmos into our skies. *WHO* could this majestic being with grandeur wings and holding a rod-spear type instrument possibly be? WHO was the white cherub escorting? *Who or what* could it possibly be?
It suddenly dawned on me with great surprise, (I didn't know how I knew, I just knew), that I was right in saying that whatever or whoever this *Being* is, it is definitely holding some type of spear-sword or rod-type instrument behind the white limbless cherub creature. So again, without knowing how I knew, I just knew it was an ***'Archangel'***.
So I started to look up information and do some research on which 'Archangel may have carried sword or spear', and here is what I found:

Archangel Michael
"Archangel Michael is considered by most of the monotheistic religions, such as Christianity, Islam and Judaism to be the *most powerful* of all other Archangels. According to the Greek Fathers and the Roman Liturgy, Archangel Michael is considered to be one of the '*Actual Cherubim*' who stood before the Gate of Paradise with a 'flaming sword' that was bathed in holy light and illuminated everything in its path".
"HE is the Archangel of battle and defender of Heaven and has a unique role as God's messenger in critical times of history and salvation: He appeared to Moses as the fire in the burning bush, he appeared to have rescued Daniel from the lion's den, and according to Christians, he was the angel who informed Mary of her approaching death. In the book of Revelation, Michael is said to be the angel who came down from the heavens, having the key of the abyss and a great chain in his hand. He is always depicted with conspicuous and very grand wings, and always carrying an unsheathed sword, shield or spear, and his skin is described as the colour of copper which 'radiates and glows' from being in the presence of God".
(See photos, PAGES 34-37)

Wow! I was speechless and I was dumbfounded! I just couldn't wrap my mind around all this information I was gathering, as I was trying to process an understanding of what I had captured in my camera, and the images I was seeing with my MFT Method. I was shocked, surprised, amazed, in disbelief, you name it every single emotion was going through my mind. And of course, my tingling sensations were constantly with me on the top of my head.
I finally came to accept that these 'tingling head sensations' were confirmation messages from the Higher Divine, and that I was being *guided*. When finally I began to *accept* these strange feelings and 'thoughts', I began to trust my intuition, I began to trust my gut-feelings and my inner-self, and that is when my research started to *soar*. I began to better understand orbs, and I began to classify orbs into categories and groups, which has never been done before.

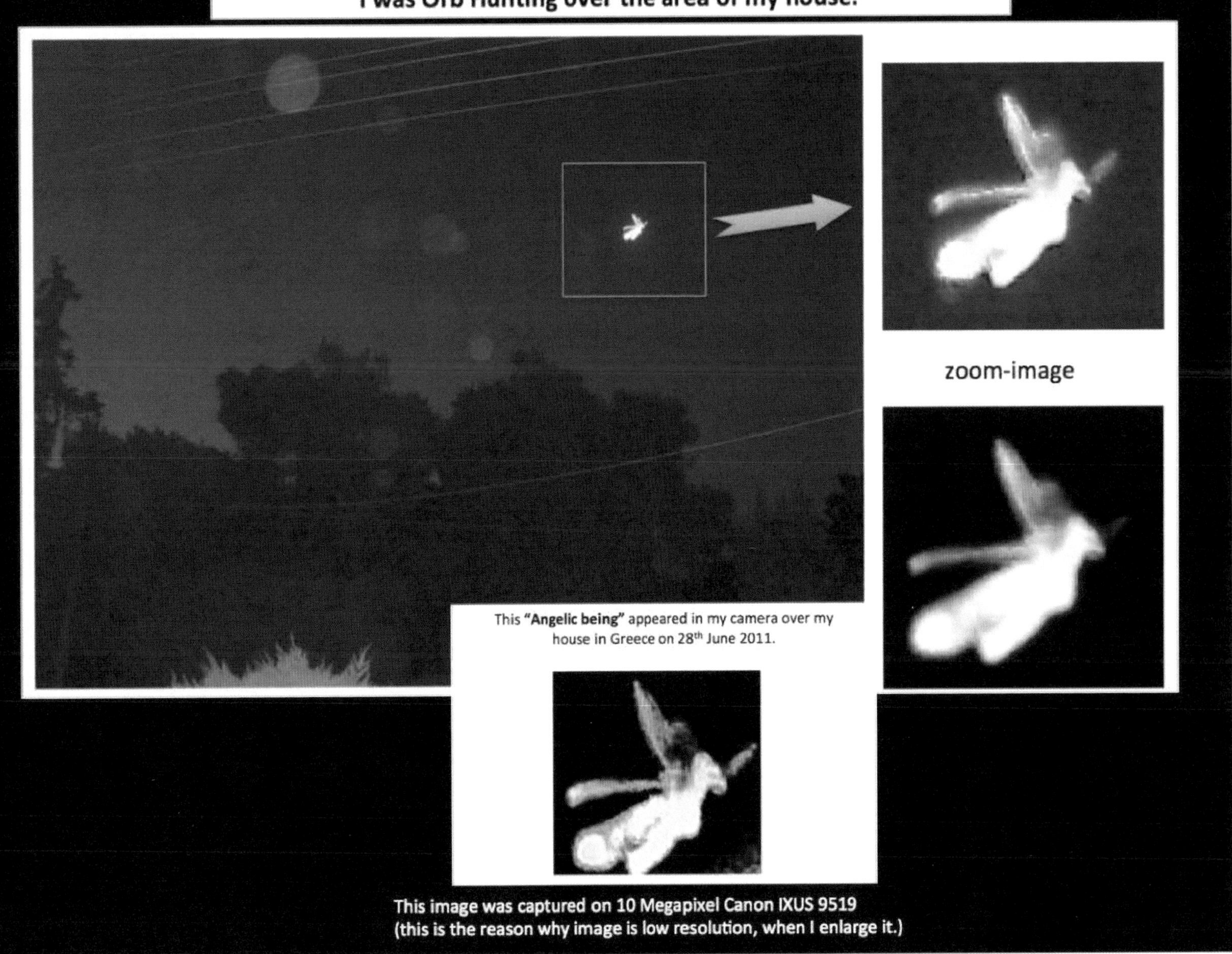
I captured this photo in Athens, Greece, on 28th June 2011, when
I was Orb Hunting over the area of my house.
zoom-image
This "Angelic being" appeared in my camera over my
house in Greece on 28th June 2011.
This image was captured on 10 Megapixel Canon IXUS 9519
(this is the reason why image is low resolution, when I enlarge it.)

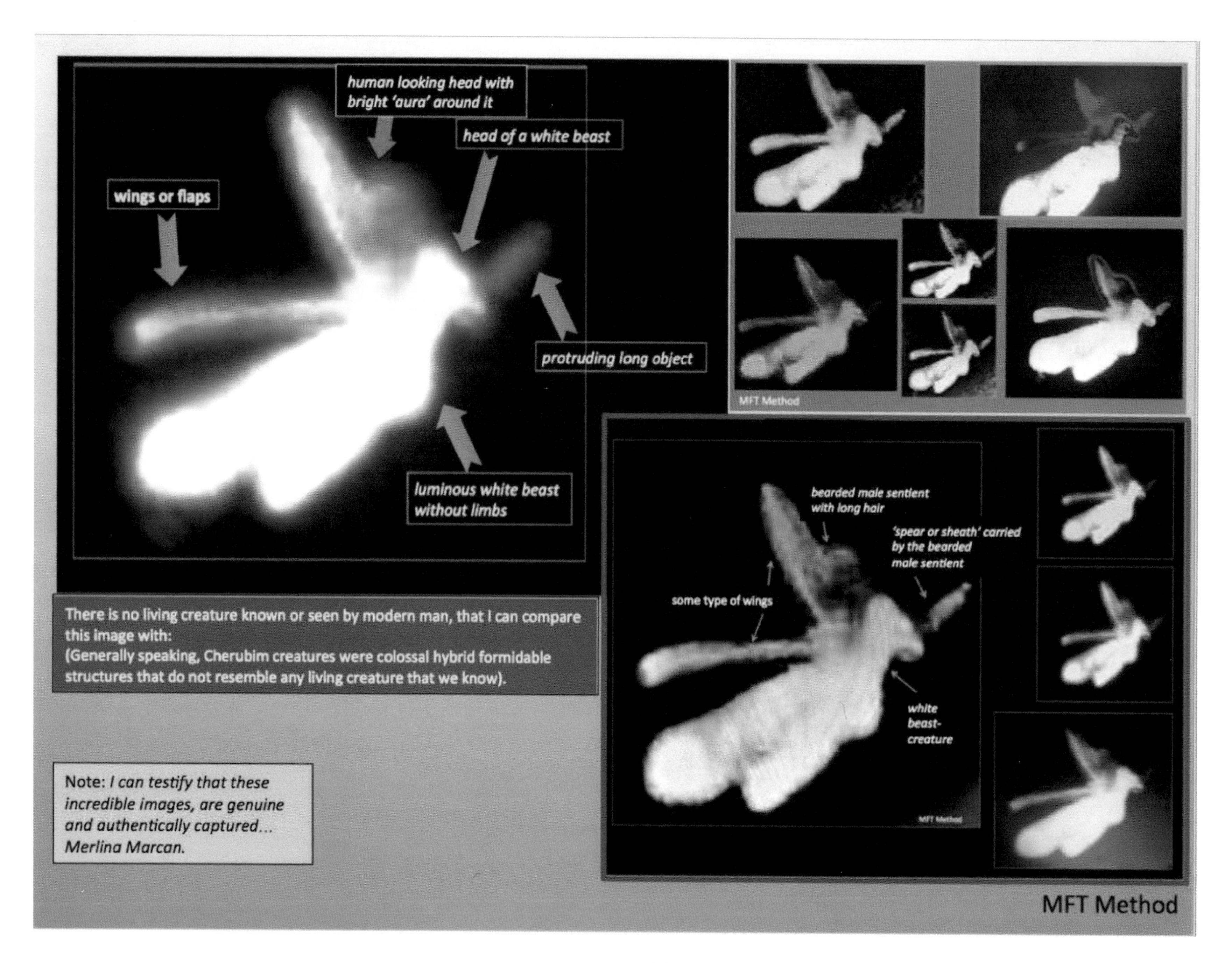
human looking head with bright 'aura' around it
head of a white beast
wings or flaps
protruding long object
luminous white beast without limbs
There is no living creature known or seen by modern man, that I can compare this image with:
(Generally speaking, Cherubim creatures were colossal hybrid formidable structures that do not resemble any living creature that we know).
Note: I can testify that these incredible images, are genuine and authentically captured... Merlina Marcan.
MFT Method
bearded male sentient with long hair
'spear or sheath' carried by the bearded male sentient
some type of wings
white beast-creature
MFT Method

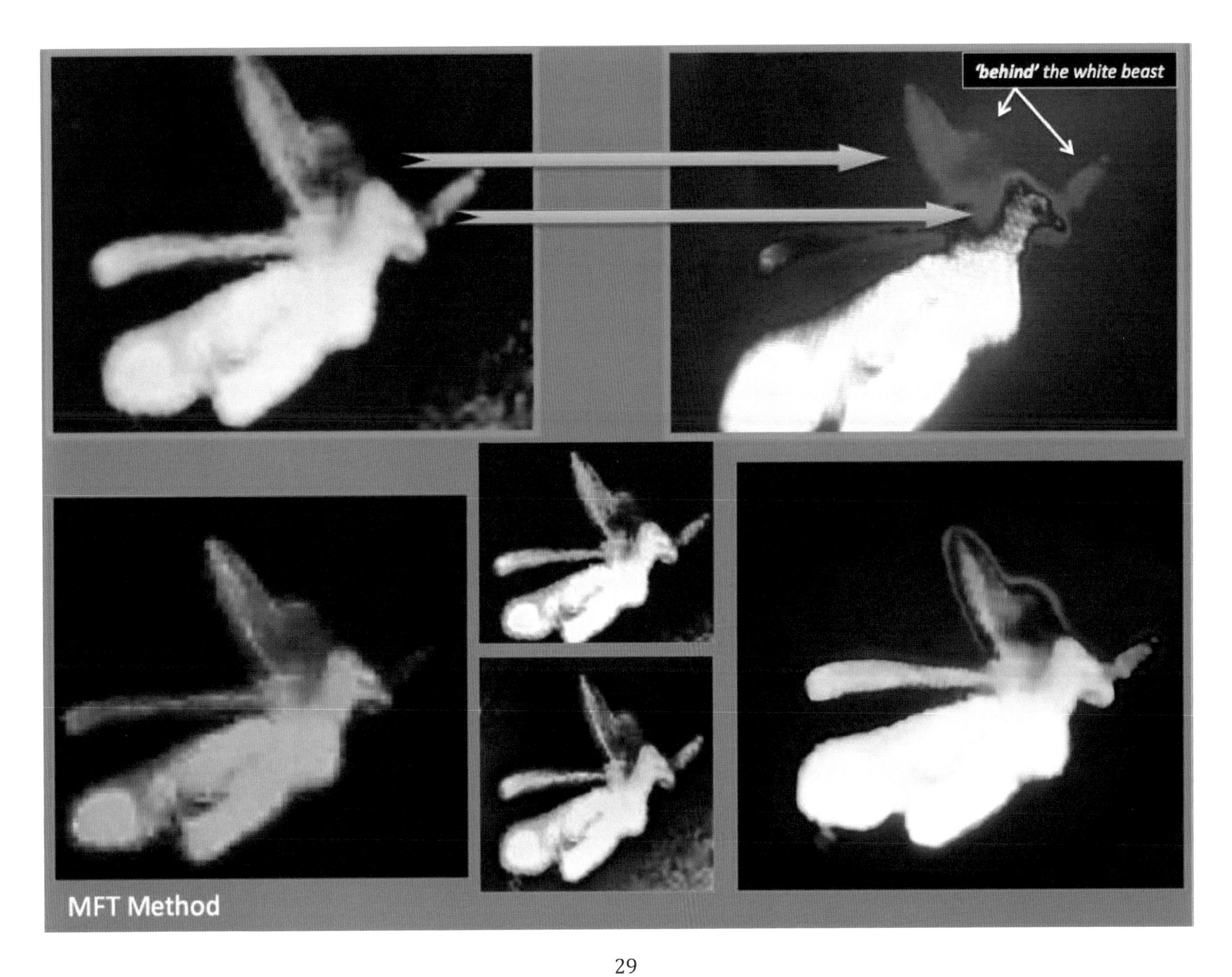
'behind' the white beast
MFT Method

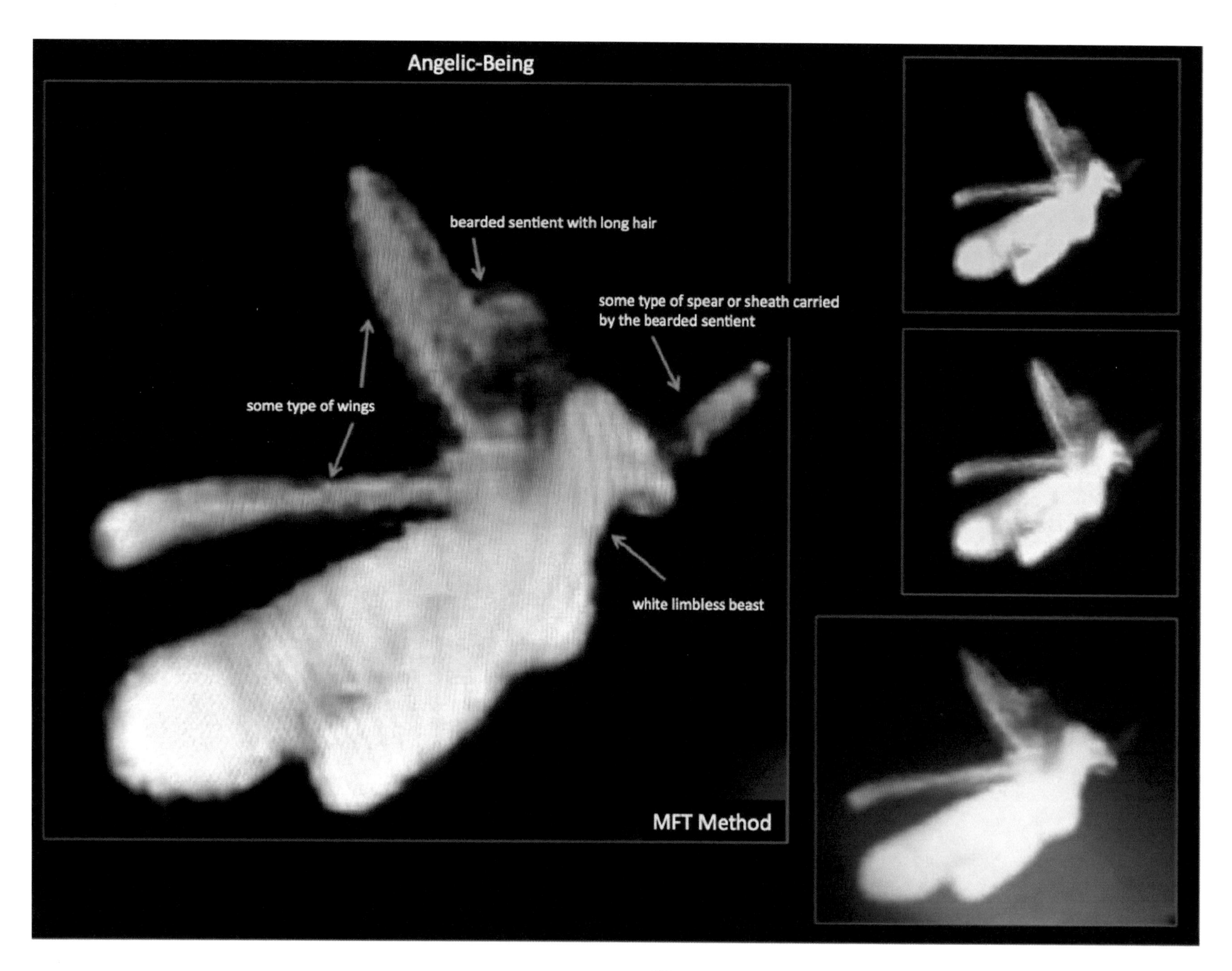
Angelic-Being
bearded sentient with long hair
some type of spear or sheath carried by the bearded sentient
some type of wings
white limbless beast
MFT Method

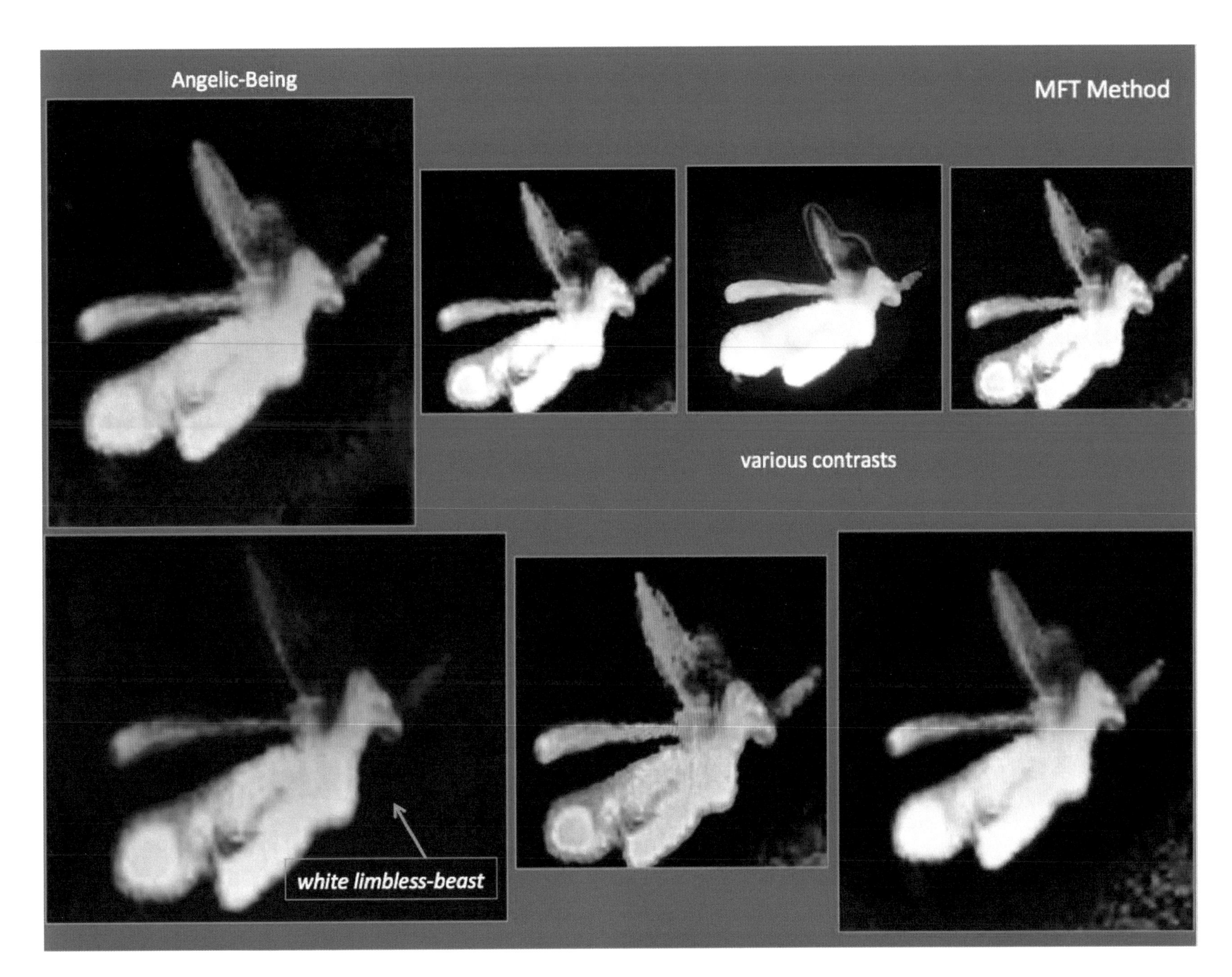
Angelic-Being
MFT Method
various contrasts
white limbless-beast

Close up view of the "dark-bearded" male sentient

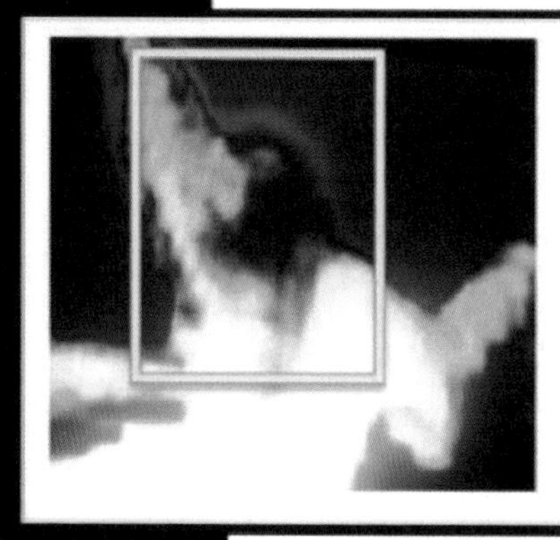

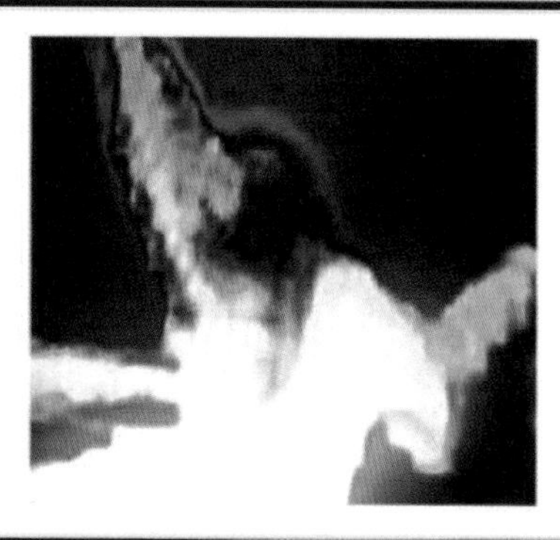

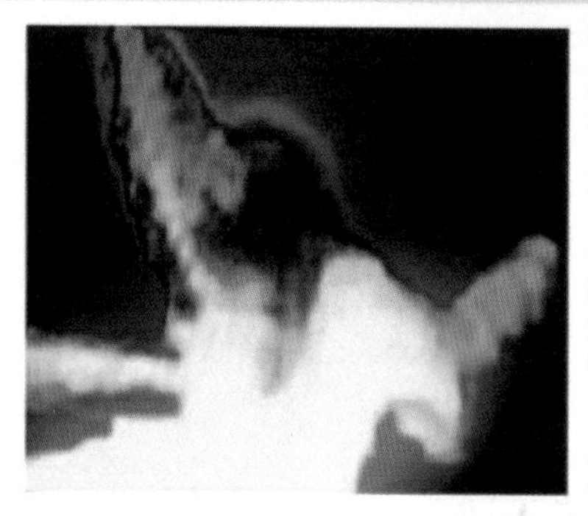

- Please note that even the 'halo-aura' around the head of this bearded sentient is a natural occurrence within the camera.

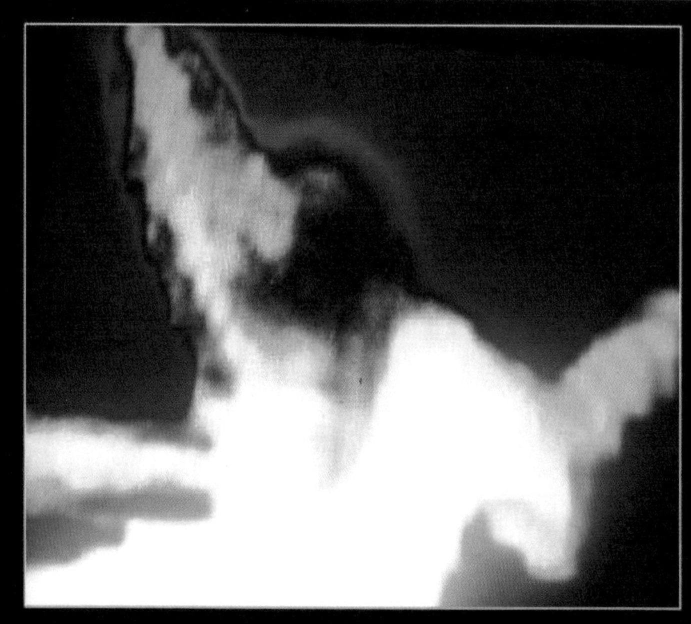

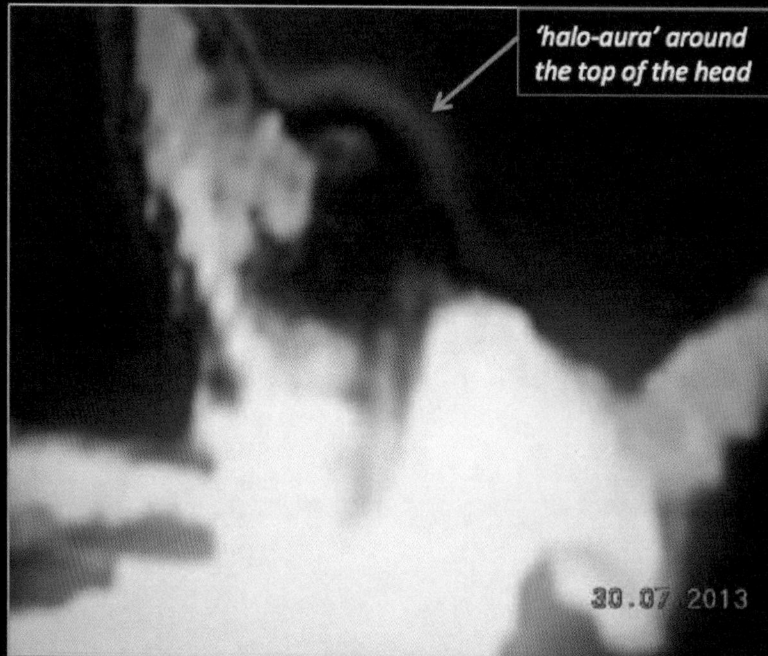

MFT Method

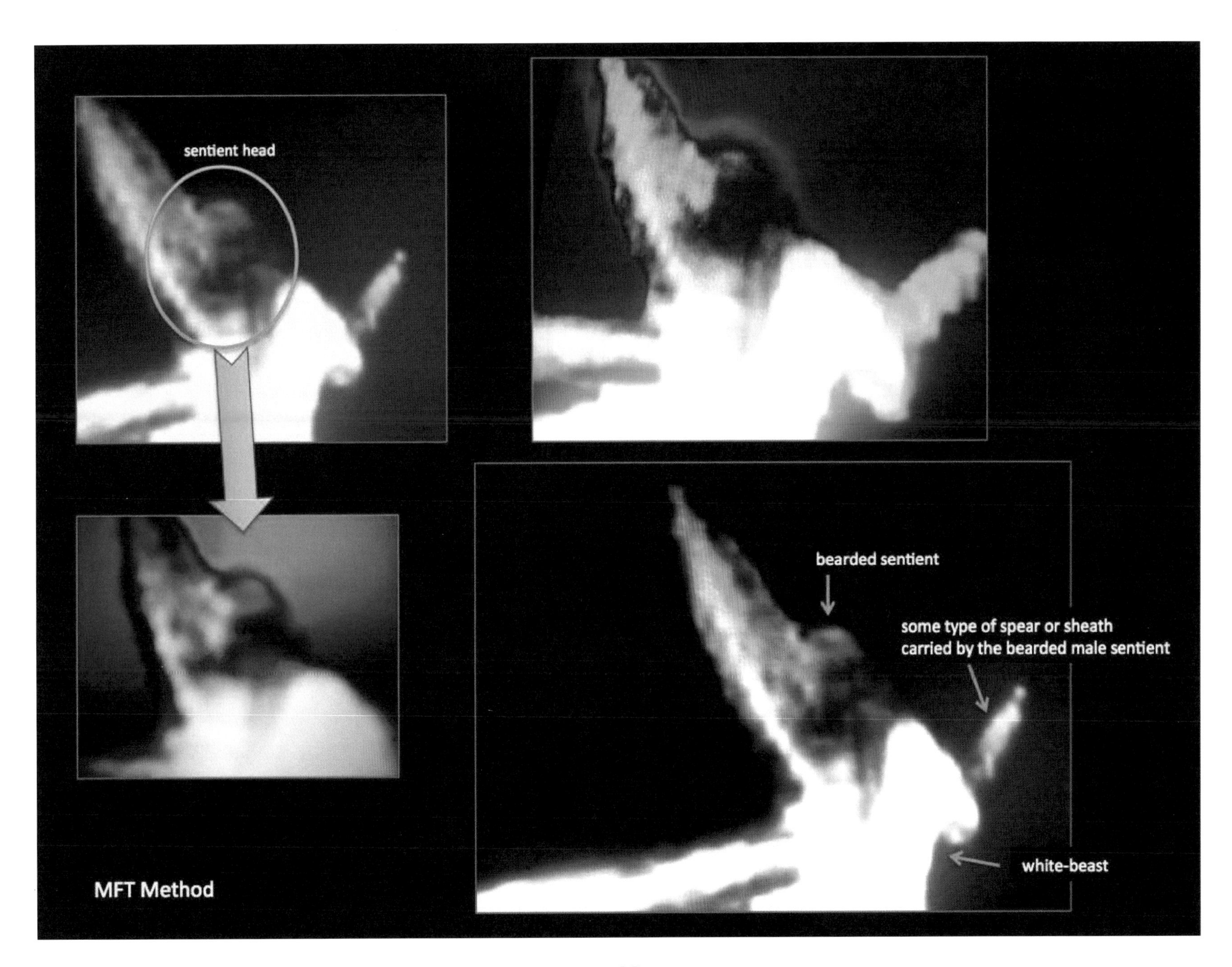
sentient head
bearded sentient
some type of spear or sheath
carried by the bearded male sentient
white-beast
MFT Method

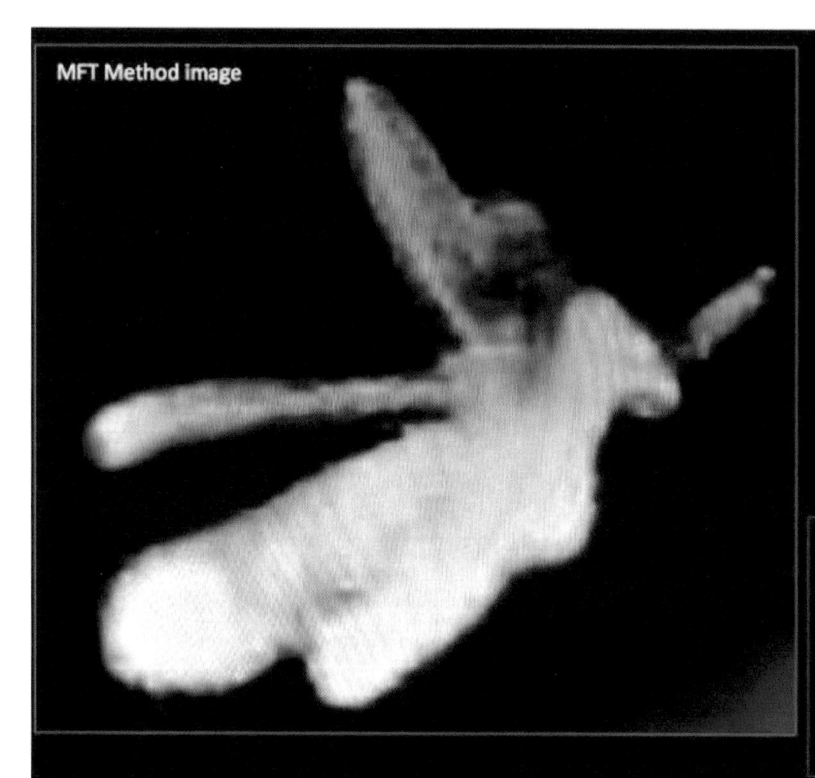

"Cherubim" in ancient times were always depicted with human heads on bodies of beasts and animals such as lions, bulls, eagles, or even a mixture of all.
For comparison: I have added some ancient stone carvings of 'cherubim' from ancient Greece, Assyrian, Babylonian, and Mesopotamian composite creatures.

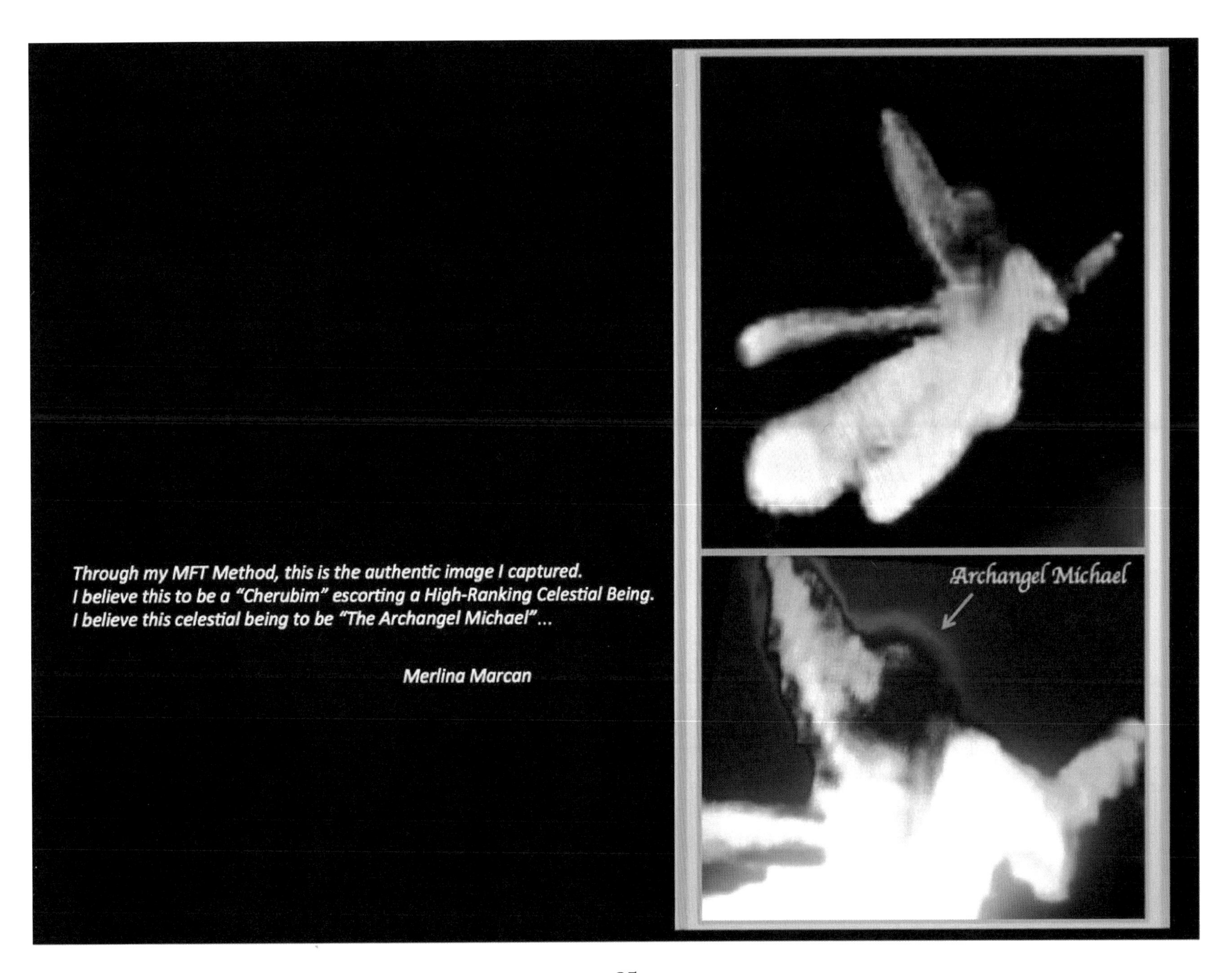
Through my MFT Method, this is the authentic image I captured.
I believe this to be a "Cherubim" escorting a High-Ranking Celestial Being.
I believe this celestial being to be "The Archangel Michael"...
Merlina Marcan
Archangel Michael

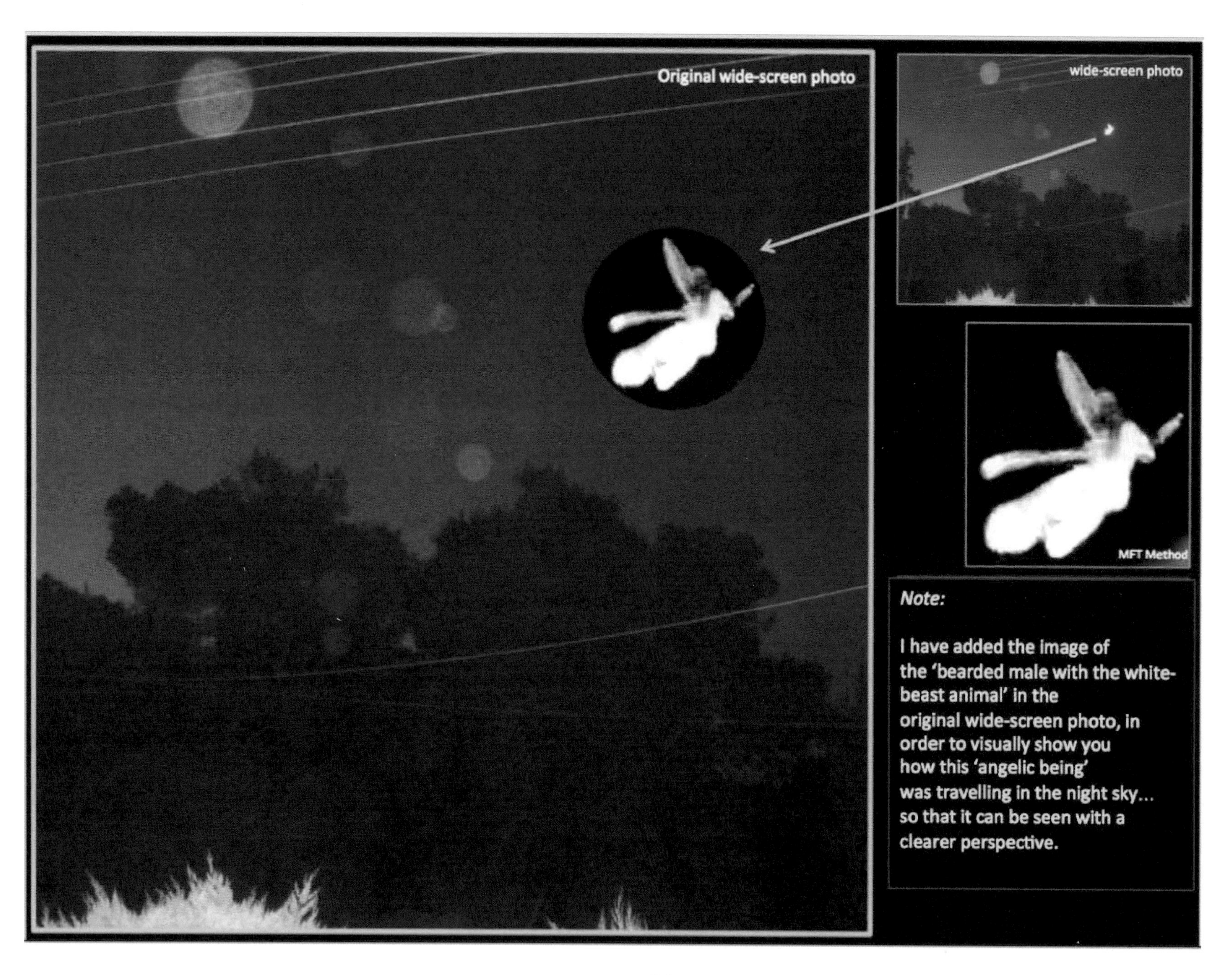
Original wide-screen photo
wide-screen photo
MFT Method
Note:
I have added the image of
the 'bearded male with the white-
beast animal' in the
original wide-screen photo, in
order to visually show you
how this 'angelic being'
was travelling in the night sky...
so that it can be seen with a
clearer perspective.

Here are some examples of icons and artist portrayals of "Archangel Michael" that can be found around the world. Archangel Michael is always portrayed with Grandeur Wings and he is always carrying a Sword, Rod, Spear, or Sheath.

Archangel Michael Monastery, Symi Island, Greece. Panormitis Icon.

Archangel Michael, 17th century icon by the Greek Icon painter Antonios Mytaras.

Archangel Michael in bronze, standing on top of The Castel Sant'Angelo, Rome, Italy..(1753)

Archangel Michael statue at St Michael's Church, Vienna, Austria. By Lorenzo Mattielli (1724)

Archangel Michael, is the patron saint of Kiev. This monument can be found in Independence Square.

Archangel Michael Fountain, on Boulevard Saint-Michel, Paris, France.

Author's note:

To some people, being '*guided from the heart*' might not seem to be a firm basis for scientific criteria findings. And I agree! But I am not basing my research on just my gut-instincts and heart alone. Actually, it is quite the opposite, because all my research is based on real, true and authentic visual evidences. The fact that I truly believe in my heart, that *I have been guided by a 'higher celestial presence' to get to where I am today, and write this book*, should ***not*** take away the impact of my work by those who do not believe in Higher presences, or Higher Beings, because ***'Logic and Truth'* are not the same thing!**

In modern world science, what is considered valid science, most often depends on the prevailing or current scientific paradigm, because very often skeptics are hesitant to accept new findings. But may I remind you, that some of our greatest inventions and greatest discoveries have come through the 'gut-feelings' of some of those brave pioneers that trusted their inner-self-intuition and *dared* to explore and break down existing boundaries, often like thunderbolt ideas, as if they had guidance from above.

Some people do not have a 'Nature of Awareness' and it is unfortunate, that in our modern conventional world, ***"some people still believe that if we cannot see it, or we cannot touch it, smell it, taste it, define it, explain it or understand it, then it doesn't exist".*** And much too often, simply because we cannot understand, or we cannot find an appropriate solution (or explanation to something that we can't wrap our minds around), people will often quickly jump to conclusions that it cannot be true or real, and therefore it *cannot* exist.
Please allow me here to give what I think is a good example of what I have just said, so I can put things into a better perspective. Let's talk about our '*minds*' for example…

Do we have sufficient knowledge of what our '*mind*' truly is? We assume that our mind is *somewhere* in the area inside our brain, even though we truly do not know where that area is. Our mind has no substance, has no weight (it is weightless), we cannot see it, smell it or touch it, it cannot be measured and it has no boundaries of space or time, and science has not managed so far to understand it. But we know our mind exists somewhere in the brain, because we use our mind every day! We use it to make our everyday choices, we use our mind to have *thoughts*, and we use our minds to travel to *past* memories, *present* happenings and *future* visions. We even use our mind to travel to ethereal virtual realities, such as in our *dreams*.

So overall, we cannot see, touch, smell, taste, define, explain or understand our MIND, does this then mean that it does not exist? For if this were the case, then we would have to also put our SOUL in the same categorization, for it *too* has the same properties. The Mind and Soul are connected, but we cannot prove this hypothesis because we cannot truly understand them.

Putting it in layman's terms:
*As long as the 'mind' is somewhere inside our brain, and the 'soul' is somewhere inside our body, we are classified as '**living beings**'...*
*But once the mind evacuates the brain, and the soul evacuates the body, we are classified as a '**lifeless corpse**'.*
For without mind and soul, we do not exist! End of story!

Mind and Soul are the two main ingredients for the recipe of 'Consciousness'. Consciousness is the basic element for all life, for without it there would be no real time and no real existence, and therefore matter would linger in an undetermined state of nothingness.
There are many things in this world that we do not understand, there are many things in this world that we cannot explain, conceptualize or apprehend, but they '*do*' exist.

And here I would like to state that:

"Sentient Beings from other dimensions '*do*' exist and can travel into our world".
I am writing this book, to give everyone the chance to see for themselves, sentient beings from other realities that are visiting our Earth, with real and authentic visual evidence. For many people, concept ideas such as *Alien-travellers and Spirit-travellers* are impossibilities and unbelievable concepts because they cannot understand or explain such phenomena. This is the reason that I am writing this book, which has accumulated from my nine-year study and research work, and I WOULD LIKE TO SHARE SOME OF MY FINDINGS.

Through my cutting-edge discovery, I have been able to capture images of Sentient Activity inside the Orb Phenomena, and to process an understanding of the Hierarchy Levels that can be found in these phenomena. I believe that Orbs are becoming more and more prominent to us, and they are becoming more and more conspicuous to us, wanting to communicate with us, in order to open up our eyes (and minds), and make us aware by seeing the true purpose of our existence, and to gain some understanding of what is to come. Perhaps this is slowly leading to the new beginning of a *NEW ERA for us here on Earth!*
"Our human destiny, lies within the exploration of these invisible unknown phenomena"...Merlina Marcan

QUOTE:
"The day science begins to study non-physical phenomena, it will make more progress in one decade than in all previous centuries of its existence"...unquote

" *If you want to find the secrets of the universe, think in terms of energy, frequency and vibration"*...unquote

Nikola Tesla

Part 6
Orbs…they select & choose 'where and when' to appear

Please Note:

I often hear some people say that in order to capture orbs, you have to visit graveyards, cemeteries, haunted old houses where people were killed, or old psychiatric and mental asylums where people were tortured etc., etc.

*Wrong, wrong, wrong! Because most of the time, orbs love to be around '**living people**' at everyday happy events, they love to interact with us and they love to watch us in our everyday activities.*
*Whereas if you go to a haunted house, or to infamous places with history of deaths and torture that had taken place, your chances of capturing **The Happy Bright and Glowing Orbs will be minimized**, and chances are that these infamous places will most probably be attracted by the low frequency earth bound entities, which many people call 'ghosts'. Ghosts are the entities that have died but have not yet gone to 'the light'. These are the lost souls that have stayed to linger on the Earth's outer platform not really knowing where to go. They linger somewhere between Earth and the Low Astral Planes where they are trapped.*

I do not believe that ghosts or 'low earth-bound entities' can use the orb phenomena as a means of transportation, because they have a 'denser atomic composition' as compared to the lighter sentient beings that can be seen travelling in orbs. Ghosts cannot travel in orbs because they have not (or will not), willingly shed their dense structure to transfer into lighter beings or 'beings of the light'.

*We will see later in these dimensions, that we will find very strict '**Hierarchy Levels**', and I will show you that orbs can be categorized in specific '**Rank Levels**'. Higher dimensions are unlike our Earthly dimension where good, bad and evil people can congregate and mix together. **In the Higher Dimensions, the good, the bad and the evil are Strictly Segregated and Divided.** In these other realities, it is not a matter of 'opposites attract' as we humans often use this expression, but instead, it is a matter of 'like attracts like' and they will always follow specific regimes or specific 'rank levels'.*

With the revolution in digital cameras since say the mid 1990's, suddenly and instantaneously all around the globe, people from different countries, different backgrounds, different age-groups, different religions and different cultures where capturing and witnessing a plethora of strange unknown phenomena in their photos, that they named as 'orbs' because of their roundish shapes. This was to bring about a controversy debate as to what these round images appearing in photos could possibly be?

The controversial debate was 'dust' versus 'spiritual realms'.
People started to question the appearance of these images. Questions such as:

a) Could they be angels?
b) Could they be our guardian angels that are watching over us?
c) Could they be spirit guides to witness our physical life-path and actions?
d) Could they be our loved ones who have passed over to the other side, and have come to enjoy and rejoice with us?
e) Could they be Celestial travellers from other worlds?
f) Could they be aliens from other planets?
g) Could they be interdimensional beings from our neighboring star systems?
h) Could they be 'dust'?
i) Could they be all of the above?

For some people, anything that seems 'intangible' to conceptualize is often easily embraced as non-existent and is often just ignored with a simple explanation that it is 'dust', and that '*all*' orbs are dust. But to say that 'all' orbs are dust is the highest form of ignorance when you 'adamantly reject something that you do not know anything about'! This of course would be as absurd as saying that *some apples are red; therefore 'all' apples are red.*
Hence, the skeptic's illogical way of thinking is to say that *some orbs are dust; therefore 'all' orbs are dust!*

There is a universal trend (and it is happening all around the world), for orbs to want to appear in our photos, because they know that our camera technology is improving, and they know that we are getting to see them in *more detail*. Orbs will often choose to appear, (or not to appear), and as a general rule they will most often choose to appear in photographs where there is a happy, content, peaceful, thoughtful and relaxed state of mind or environment. They will often appear at entertainment venues, at family get togethers, at weddings, at seminars, at healing and meditation events, at religious sites and sanctuaries where people are praying, in many homes where there is a positive calm aura, in beautiful and peaceful landscapes and nature, and just about anywhere and everywhere where there is positive energy.
In general, the appearances of bright luminous and radiant orbs are often connected with happy, or positive emotional moments.
(See photos, PAGES 46-47)

How can we possibly ignore such phenomenal images and simply say it is dust?
Images that can be refulgent bright, sometimes they are luminous and semi-transparent or thickly textured and opaque, and sometimes they can be seen penetrating solid objects.
But perhaps one of the most phenomenal things about orbs is that *they will show intelligence*, as they love to '*strategically position themselves near and around us*' (especially near the areas of our heads), and they will often do this **WHEN WE ASK THEM TO APPEAR IN OUR PHOTOS.**
(See photos, PAGES 48-51)

During nine years of my research of the orb phenomena, I have found that there is a strong bridge between 'Positive Human Emotion' and 'Bright Orbs'. The brighter an orb is, the more positive energy we can see from the people in the photo. So let's keep in mind when luminous bright orbs are present, we will experience positive energies or positive emotions.

As a universal concept, I have noticed that '***orbs often love to hover around our heads***', or to keep a close range around the areas of our heads. It is as if they are trying to make ***Mind Contact***. Perhaps to send us messages in our sub-conscious minds that may be perceived to us as 'thoughts', 'ideas' or 'feelings'. So next time you see a bright orb around a person in a photo, try noticing to see what that person is doing or feeling. Most often that person will be joyful, happy, loving, playful, content, in a meditative state, or in a peaceful and relaxed state of mind!

Could it be, that sometimes when we feel an unusual emotion or we get an electrifying or bright idea that seems to come from out of the blue, or we get an unusual gut-feeling from out of nowhere…could it be because an orb is hovering around us (which we cannot see with our naked eyes)? Could it be that it is sending us sub-conscious messages and it is inter-acting and influencing our thoughts at that moment?
(See photos, PAGES 52-53)

Orbs are very inquisitive and they enjoy our company. ***They especially have a strong connection to 'children' and 'animals'***, and can often be seen in photos hovering around them or following them. Children and animals are usually happy, playful and loving, so it is not surprising for orbs to want to be around them. I believe that many animals such as cats and dogs for example, can see these luminous light beings with their naked eyes, because of their ability to see a wider range of the spectrum than we humans have.
I also believe that ***newborn babies may have the ability to visually see orbs too***, but as they get a little older, they lose this ability. I am sure many of you have seen newborn babies, smiling or giggling and playfully interacting with 'something in the air' and trying to reach out and touch something in the air, that we can't see, but they can! I believe when newborns react this way for no reason; it is because an orb is visiting the newborn baby.
Orbs often visit children and animals to watch over them. So parents should not be concerned or alarmed if they take a photo of their children and orbs appear in the photo. Children with orbs around them are being 'watched over' and are being 'protected' by these beautiful energies of light.
(See photos, PAGES 54-57)

Orbs may also visit us at sad and difficult times to 'COMFORT US'!
They will often appear at grieving moments when we have lost a loved-one, or when we are feeling a little sad, sick, indisposed, or confused. They will often come to give us reassurance and perhaps send us 'energy waves of thought into our sub-conscious' that all will be well. ***But perhaps one of the most intriguing things that they can do is to 'PROTECT US'!***
I believe that they can sometimes intervene to 'protect us', and that CELESTIAL INTERVENSION is happening all around the world in our everyday lives, more so than we think.
As the old saying goes: 'if it's not your time to go…*it's not your time to go*! It makes sense now!
(See photos, PAGES 58-61)

'they' visit us at special & happy events:

photo credit: Michelle P (Australia)

photo credit: Antonio Jimenez (Mexico)

photo credit: Babie Thi Hatfield (USA)

photo credit: Cathy Finch (UK)

photo credit: Jacqueline White (UK)

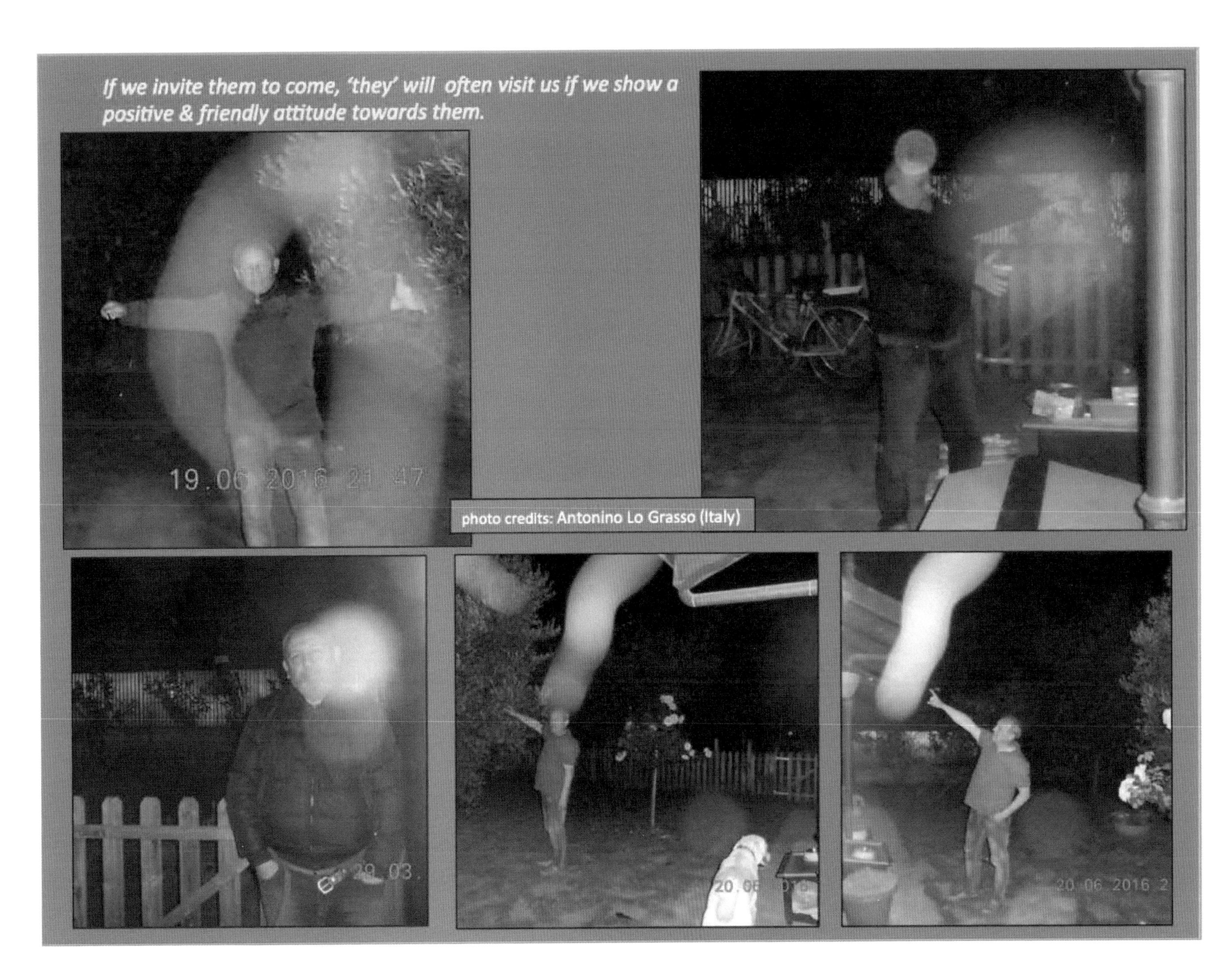
If we invite them to come, 'they' will often visit us if we show a positive & friendly attitude towards them.
19.06.2016 21:47
photo credits: Antonino Lo Grasso (Italy)
20.06.2016 2

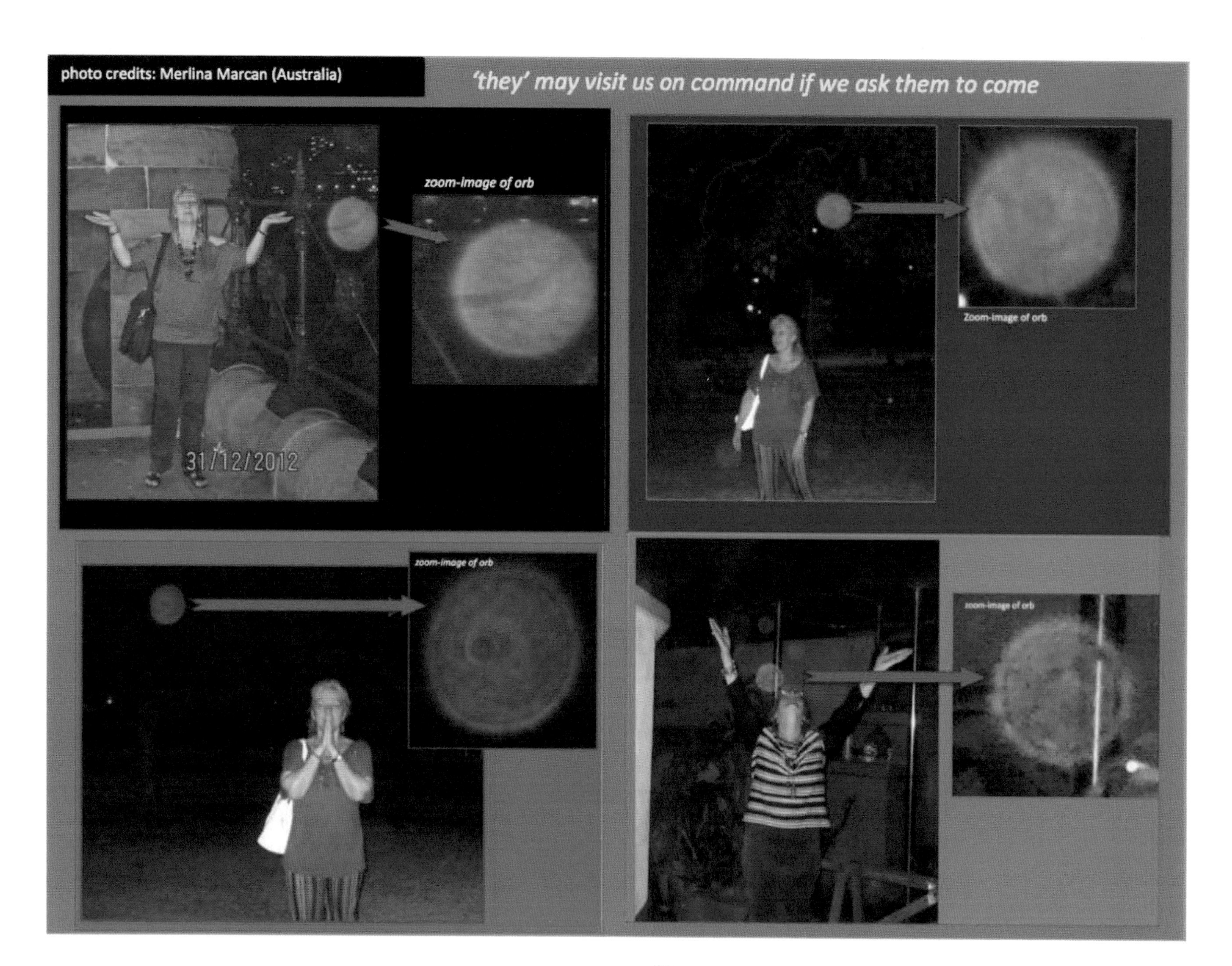
photo credits: Merlina Marcan (Australia)
'they' may visit us on command if we ask them to come
zoom-image of orb
31/12/2012
Zoom-image of orb
zoom-image of orb
zoom-image of orb

photo credits: Cathy Finch (England)

'they' may visit us on command if we ask them to come

photo credits: Diana Davatgar (Germany)
'they' may visit us on command if we ask them to come

photo credits: Peter F Kahrs (Norway)

'they' may visit us on command if we ask them to come

(c) Peter F Kahrs

(c) Peter F Kahrs

(c) Peter F Kahrs

photo credit: Cathy Finch (UK)
photo credit: Paul Greco (USA)
photo credit: Andrea Corsick (USA)
photo credit: Ricke Tan (Malaysia)
photo credit: Babie Thi Hatfield (USA)
photo credit: Cathy Finch (UK)
photo credit: James B Lindsay (USA)

'they' come to give us reassurance

photo credit: Rockette Marie (USA)

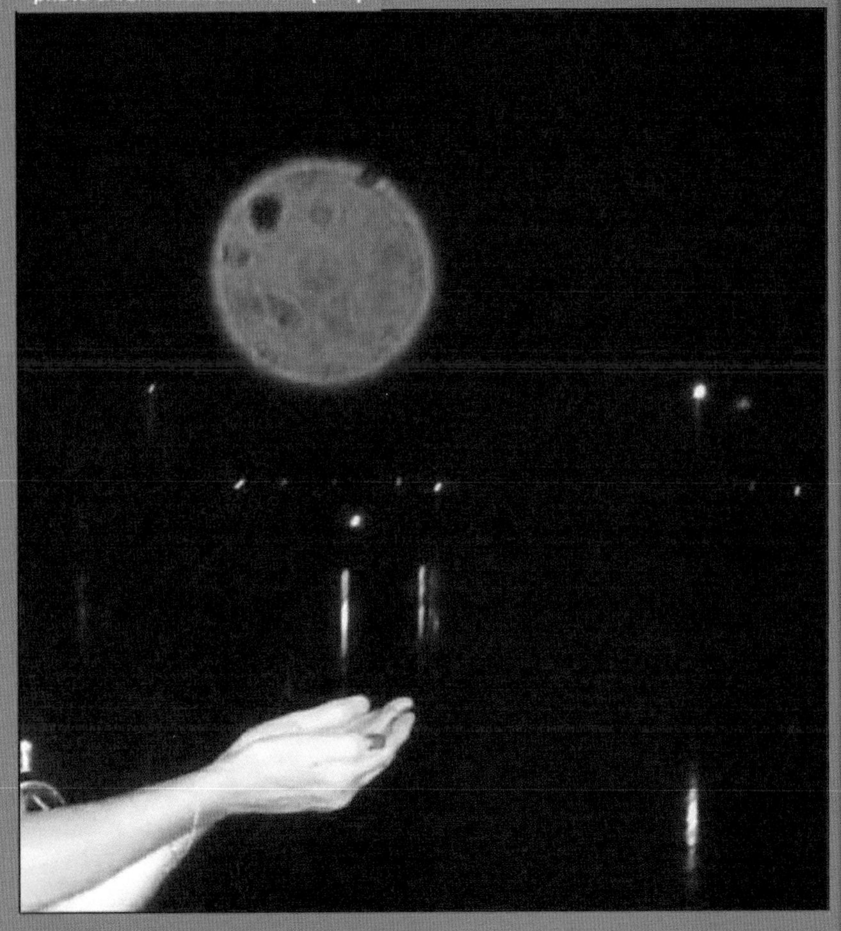

Rockette Marie:
quote: *"While asking for my mom who passed 11 years ago, she came as this blue intricate orb"*...unquote.

'they' visit us at peaceful sanctuaries

photo credit: Cathy Finch (UK)

'they' have an amazing curiosity and sense of playful humor

photo credit: Dawn Blackburn (UK)

This orb is nestled BEHIND the leaves as if it is playing 'hide & seek' with the photographer.

photo credit: Babie Thi Hatfield (USA)

This curious orb has opened up a window to get a better and closer view of the action that is happening.

Photo credit: Merlina Marcan

This playful orb appeared over my 'brother's head' because my sister-in-law and myself were playfully asking for an orb to appear on my brother's head who is an 'orb-skeptic'!

'they' love to watch over children in their everyday lives...whether it is watching television, playing, doing their homework, or resting.

photo credits: Kat J Peck (UK)

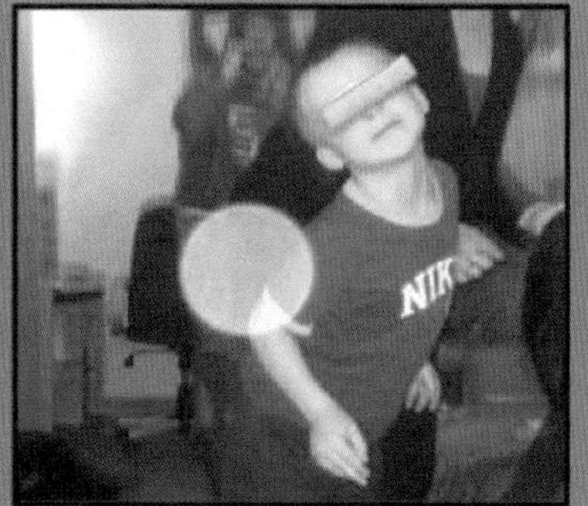

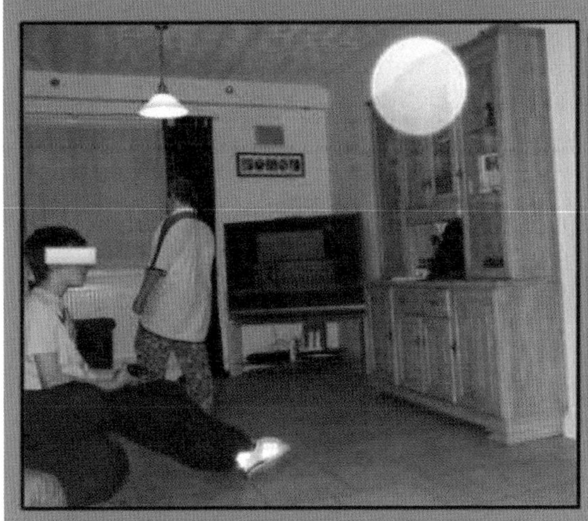

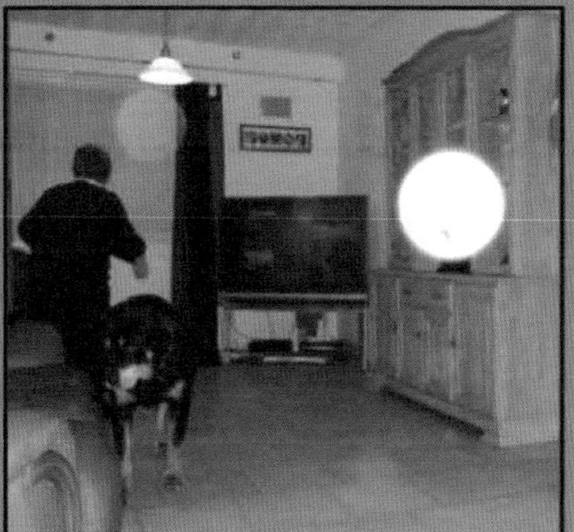

'they' love to watch over animals

photo credit: Kathy Jeffries

photo credit: Ronnie R de Wit (Netherlands)

photo credit: Melita Jelenic

photo credit: Diana Davatgar (Germany)

photo credit: Antonino Lo Grasso (Italy)

photo credit: Marion Atehsa Cyrus (Germany)

'they' love to watch over animals
quote from Karen Wilkin:
"I asked for a large light being to land on the dog! I was surprised! This is using telepathy to communicate, and yes they have proven it to me that they can hear me many times. So when they connect with you, they really start showing up, and no I do not see them, but I sense their energy. The camera picks up so much that we cannot see!"...unquote.
photo credits: Karen Wilkin (USA)

'they' will come to comfort the sick and the dying…

(this sick old dog was only hours away from passing over to the other side. The owner took many photos of her dog during these last few hours, where many white orbs could be seen hovering around the sick dog).

<u>In the last photo:</u> the dog has finally passed away (it is covered beneath the brown blanket) and a golden-yellow orb can be seen hovering over the dead dog.

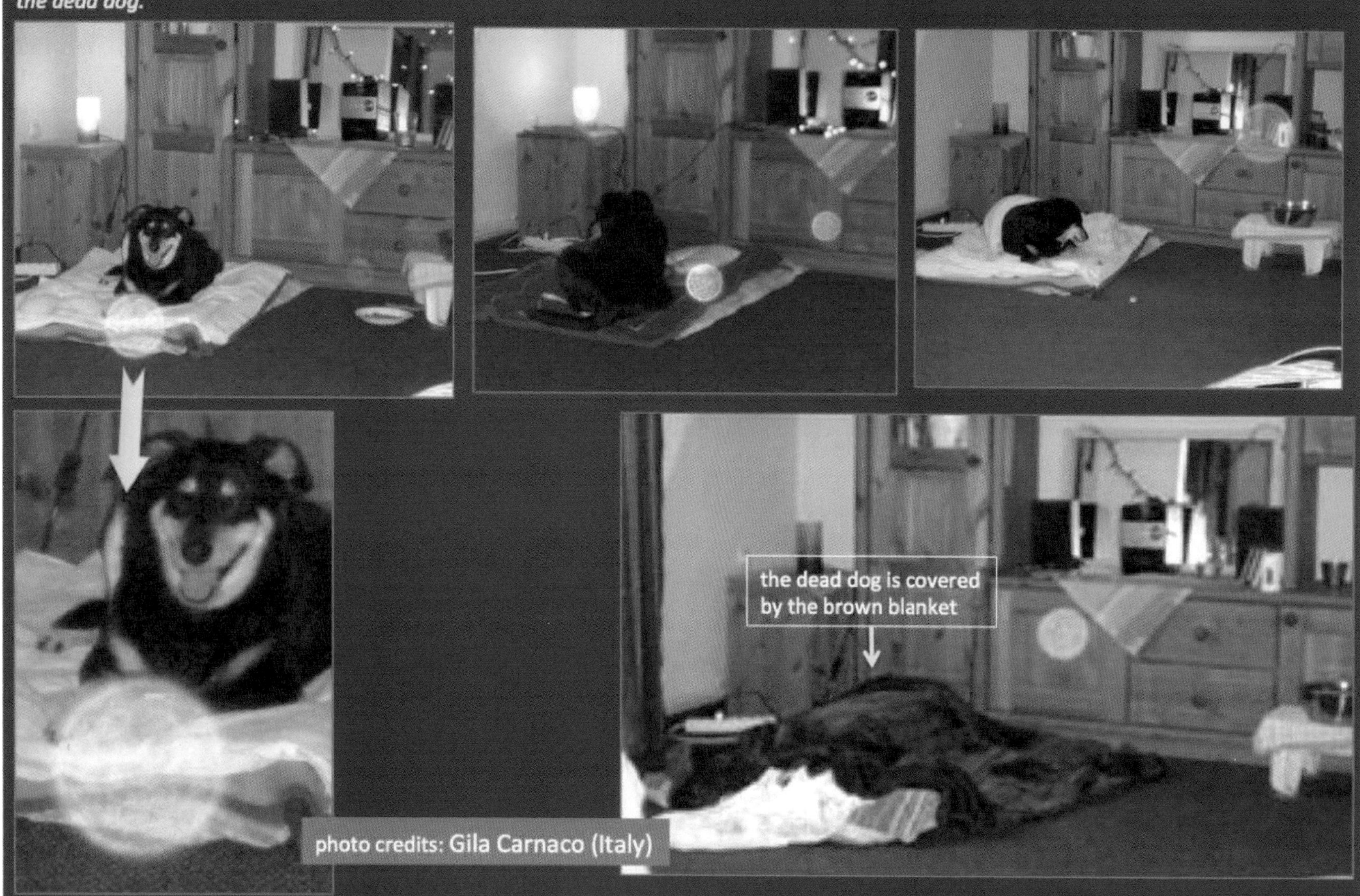

photo credits: Gila Carnaco (Italy)

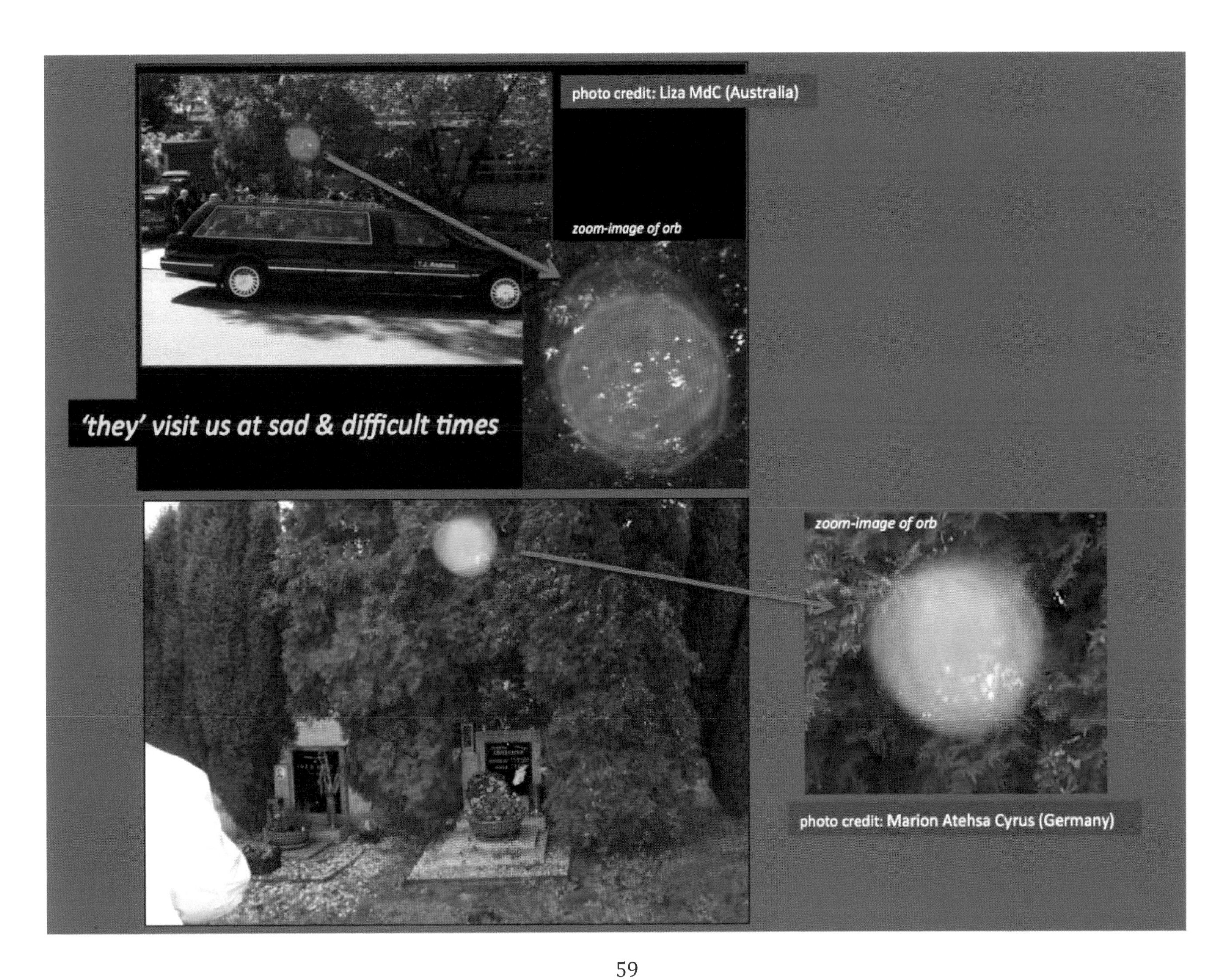
photo credit: Liza MdC (Australia)
zoom-image of orb
'they' visit us at sad & difficult times
zoom-image of orb
photo credit: Marion Atehsa Cyrus (Germany)

'they' come to comfort and guide us for our Spiritual Transition.

Lama Serpo *was a renowned Dzogchen Yogi from Bhutan, a great practitioner and a scholar.*

In his final hours of life here on Earth, this honorable Yogi was being visited by a 'semi-translucent orb around his head', perhaps comforting him, and preparing him for the next phase of his spiritual journey. (As can be seen, this orb is perfectly round in shape which means it has completely stopped in its motion to perform its mission).

Lama Serpo passed away peacefully in 2011 at the age of 97.

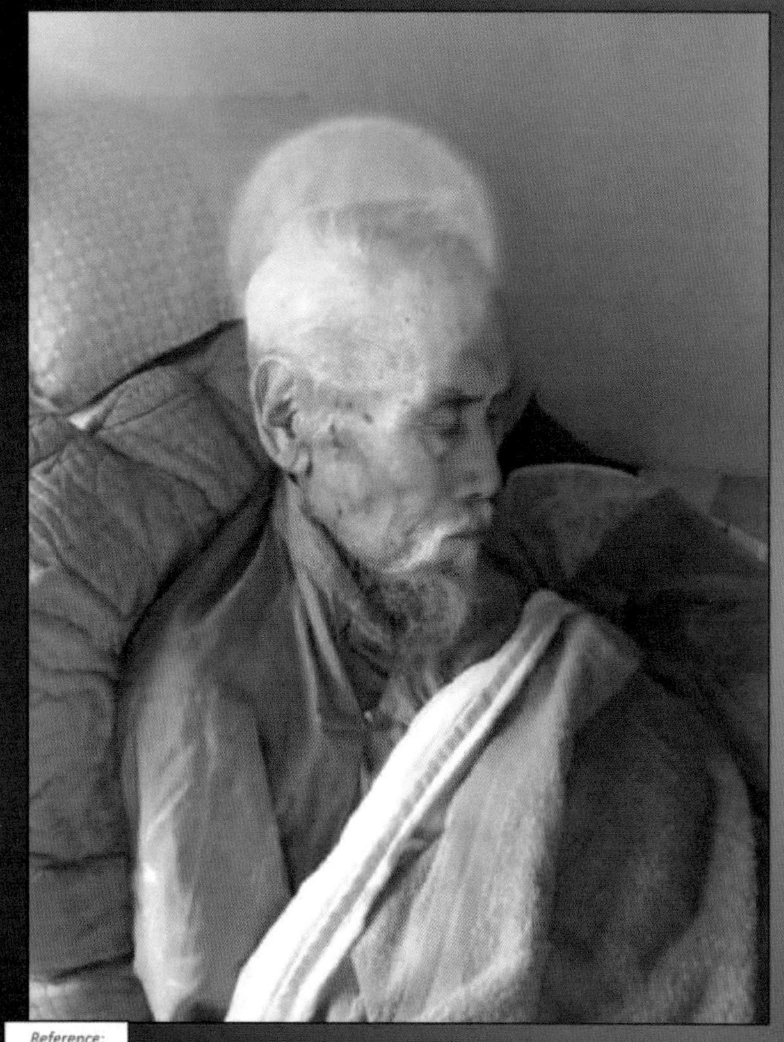

Reference:

Naljore Gi Wangchuck Dampa Jetsun Lamai Zednam Ngejung Gi Phonya Tharpa Dang Thamched Khenpai Lamten Zhey Jawa Zhungso.

Approval: From Thampa Pethey, A disciple of Lama Serpo.

Posted by Phub Dorji Wang at March 18, 2019

'they' come to protect us

photo credit: Sunshine Lanning (USA)

quote from Jayde Lanning:
*"I would like to share one of my favorite pics.
My daughter, who was 16 at the time, took this pic of her boyfriend. They were night hiking here in Colorado Springs. He was standing on a cliff and outstretched his arms. As my daughter, was snapping the photo, he said 'I hope I don't fall!' This was the resulting photo. She said she saw the light shoot out in front of her!"*...unquote.

Part 7
Colours...Size...Shapes...& Direction of orbs

Orbs have ability to change their colours, size, shape and their direction within the blink of an eye. Unlike our technology, *their technology* is perhaps hundreds of thousands of years more advanced than ours. They can *re-shape, re-size and re-colour*. They have mastered the art of shape shifting at will, and they have also mastered the art of being able to change their atomic structures at will. Orbs can be seen in *unlimited varieties of shapes and colours*. They have the amazing ability to change their colours at will, which I believe are '*Aura Shields of Protection*'. They change depending on their various 'missions' that they have to undertake. I believe that they do this for their own cosmic protection as they travel through different light, and different atmospheric pressures and conditions.

Perhaps 'different aura shields of colours' can produce 'different specific abilities for them', as for example blue auras may have different functional abilities as say from orange auras, pink auras, red auras etc. But this is not strange, if we consider how many species in our animal and insect kingdoms have this ability to change their external colours as a means for protection and survival to camouflage from predators or hostile environments. As for example, the Chameleon Lizards are quick colour-changers of the animal kingdom, rapidly altering their skin colours from greens, to yellows to reds at will. Also some fish, octopus and even many insects have this colour changing ability. So it is not surprising that *orbs* can change their aura shield colours with added thousands of years of more technology and more experience than us earthlings.
(See photos, PAGES 64-65)

Orbs and other realities that are visiting here on Earth, are not limited by boundaries of space, time, size or shape. They may appear as small as the size of a golf ball and then shape-shift in size as large as a car. They have no limitations in their shape-shifting abilities, and can quickly and easily shape-shift from one structural shape to another effortlessly, as they can phase in and out into different dimensions by changing their atomic structures.
(See photos, PAGES 66-67)

Orbs are 'Travel Capsules' or 'Travel Chambers' from other frequencies.
If we humans can grasp this concept, it will be easier to understand that they are similar in many ways to our *spaceships* that travel in our reality, the only main difference is that they are thousands of years more advanced.

Note:

- If an orb is in a complete '*Stand-Still position*' **(ZERO MOTION MODE)**, it is completely round in shape; it is generally surrounded by 'neutral-colour emissions' or cool colour emissions, such as grey or beige colours etc.

- If an orb is in '*Slight Motion*' **(SLOW HOVERING)** and is asymmetrical, it generally means that it is travelling in the direction of the warm-colour emissions (reds, orange, yellows), and it is moving away from the direction of the cool-colour emissions (blues, purples, violets).

- If an orb is in '*Super Motion*' **(EXTREME VELOCITY)**, it is usually seen as a 'beam of light', and this means it is travelling at 'unbelievable speeds'.

(See photos, PAGES 68-74)

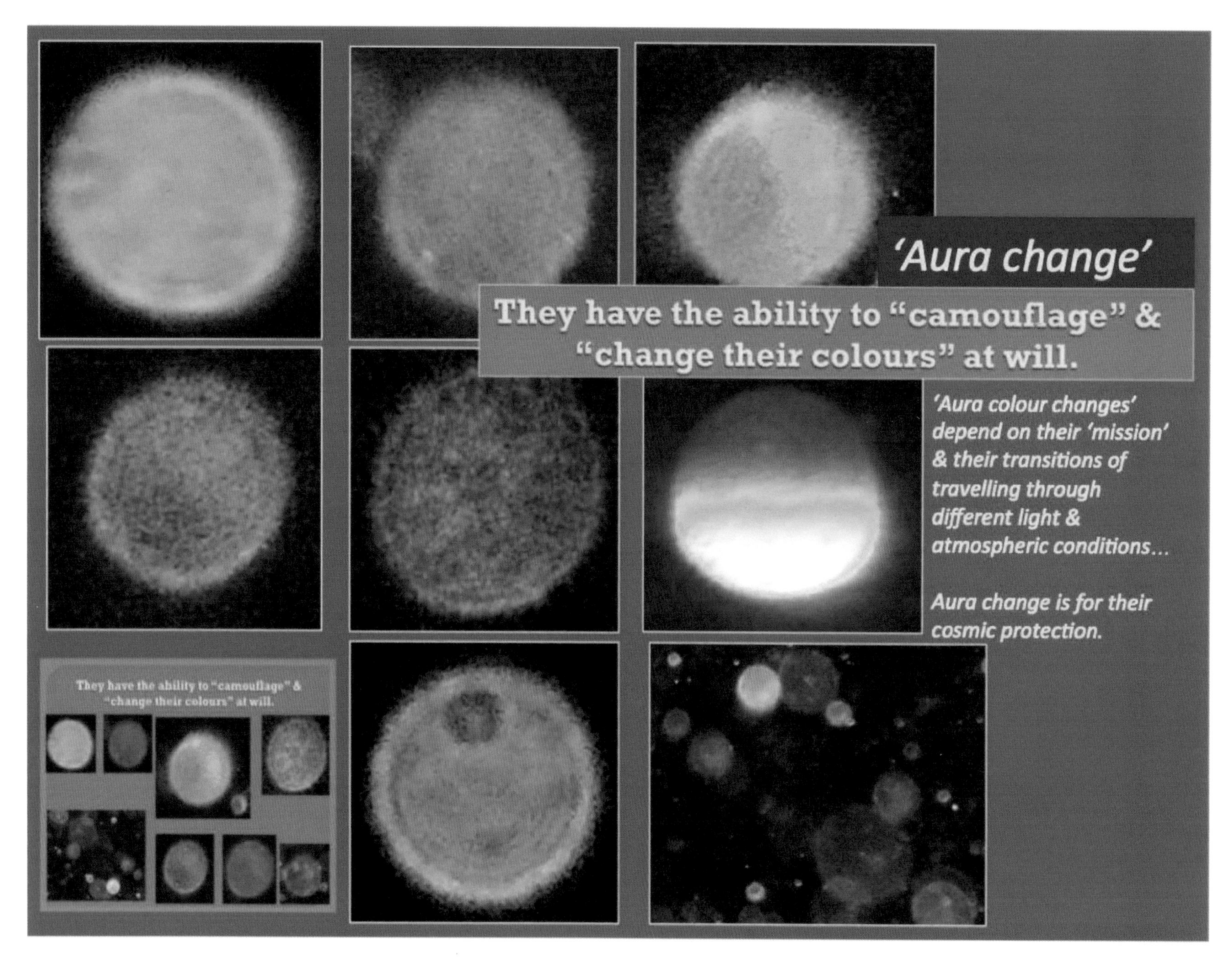
'Aura change'
They have the ability to "camouflage" &
"change their colours" at will.
'Aura colour changes' depend on their 'mission' & their transitions of travelling through different light & atmospheric conditions...
Aura change is for their cosmic protection.
They have the ability to "camouflage" &
"change their colours" at will.

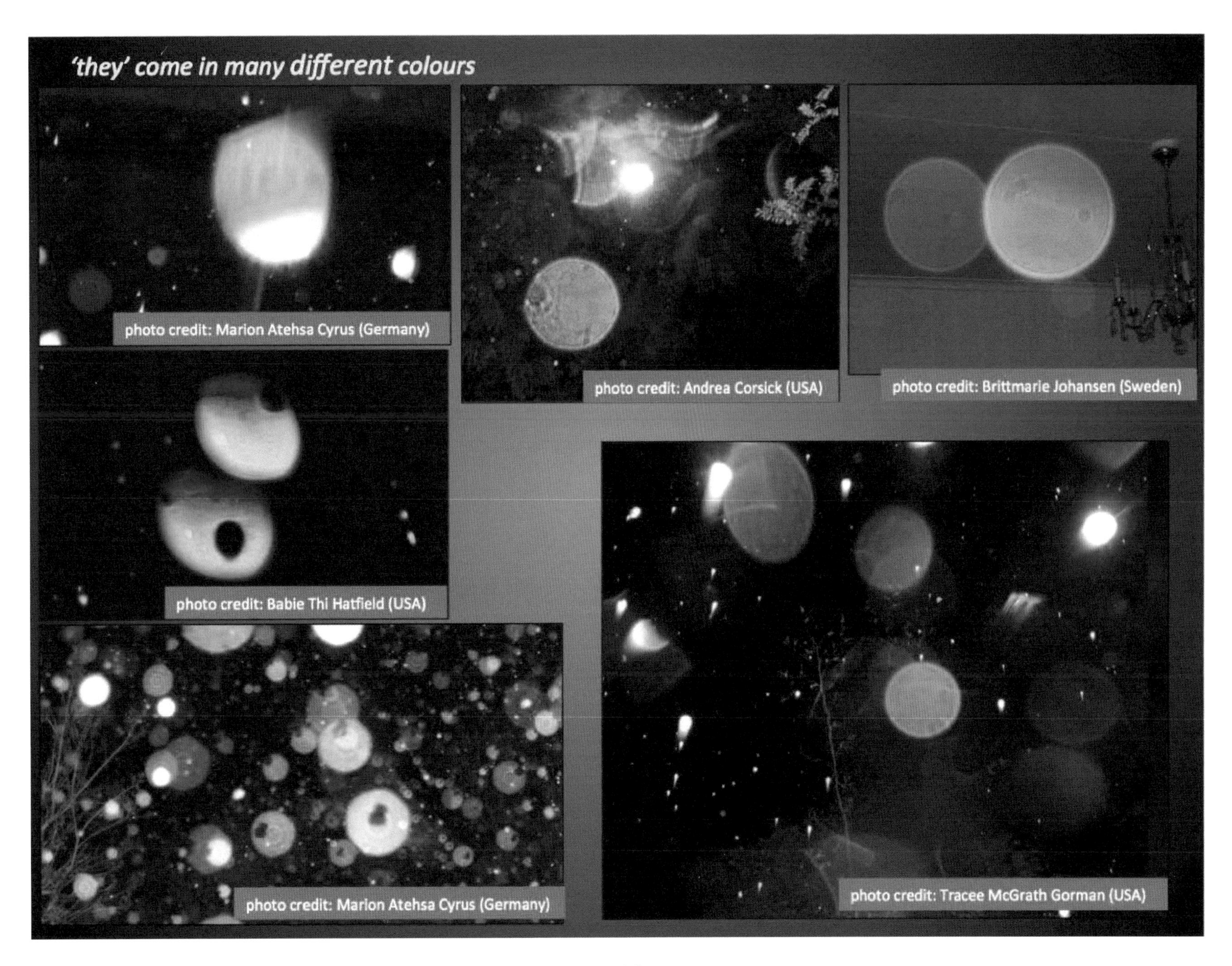
'they' come in many different colours
photo credit: Marion Atehsa Cyrus (Germany)
photo credit: Andrea Corsick (USA)
photo credit: Brittmarie Johansen (Sweden)
photo credit: Babie Thi Hatfield (USA)
photo credit: Marion Atehsa Cyrus (Germany)
photo credit: Tracee McGrath Gorman (USA)

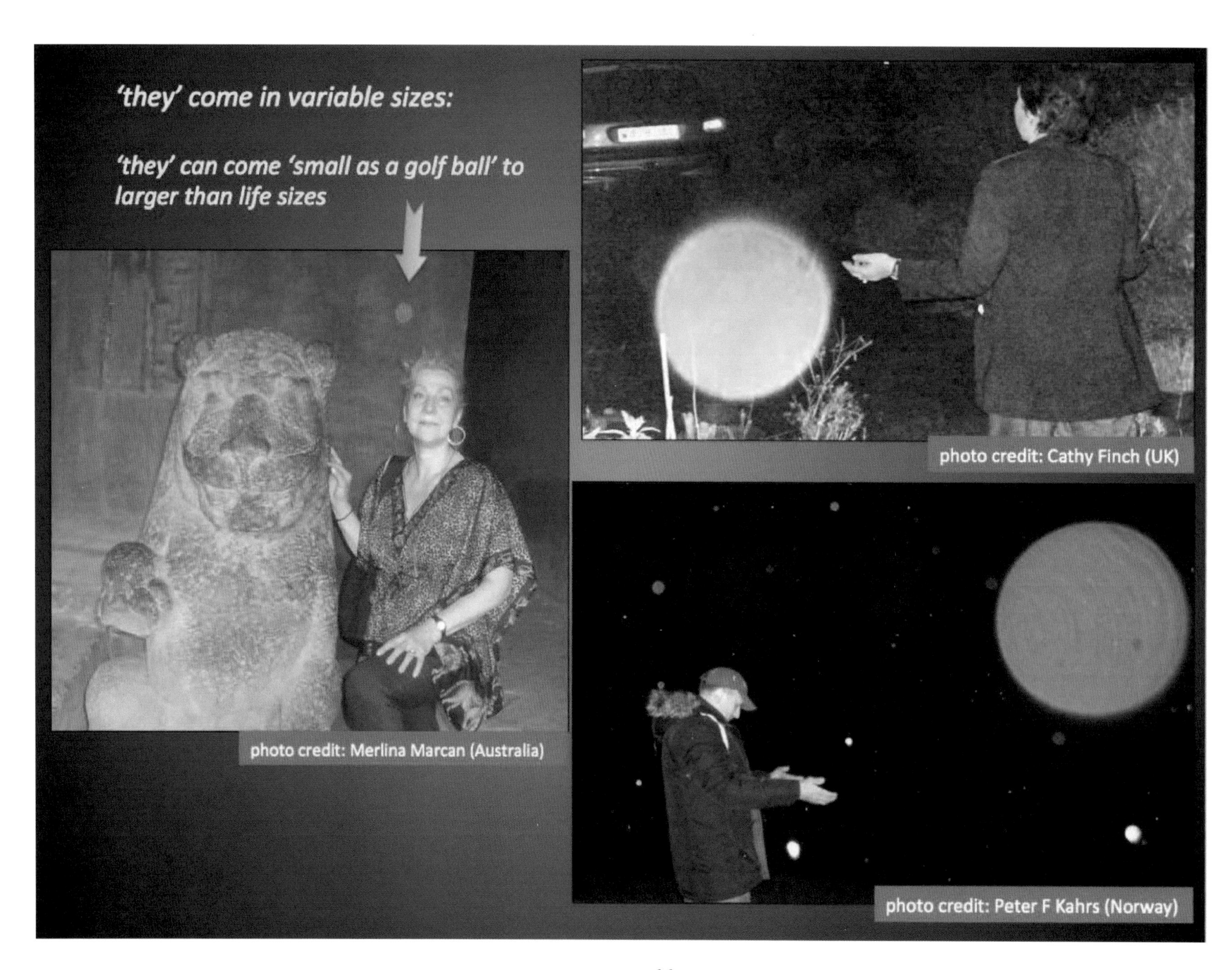
'they' come in variable sizes:
'they' can come 'small as a golf ball' to
larger than life sizes
photo credit: Cathy Finch (UK)
photo credit: Merlina Marcan (Australia)
photo credit: Peter F Kahrs (Norway)

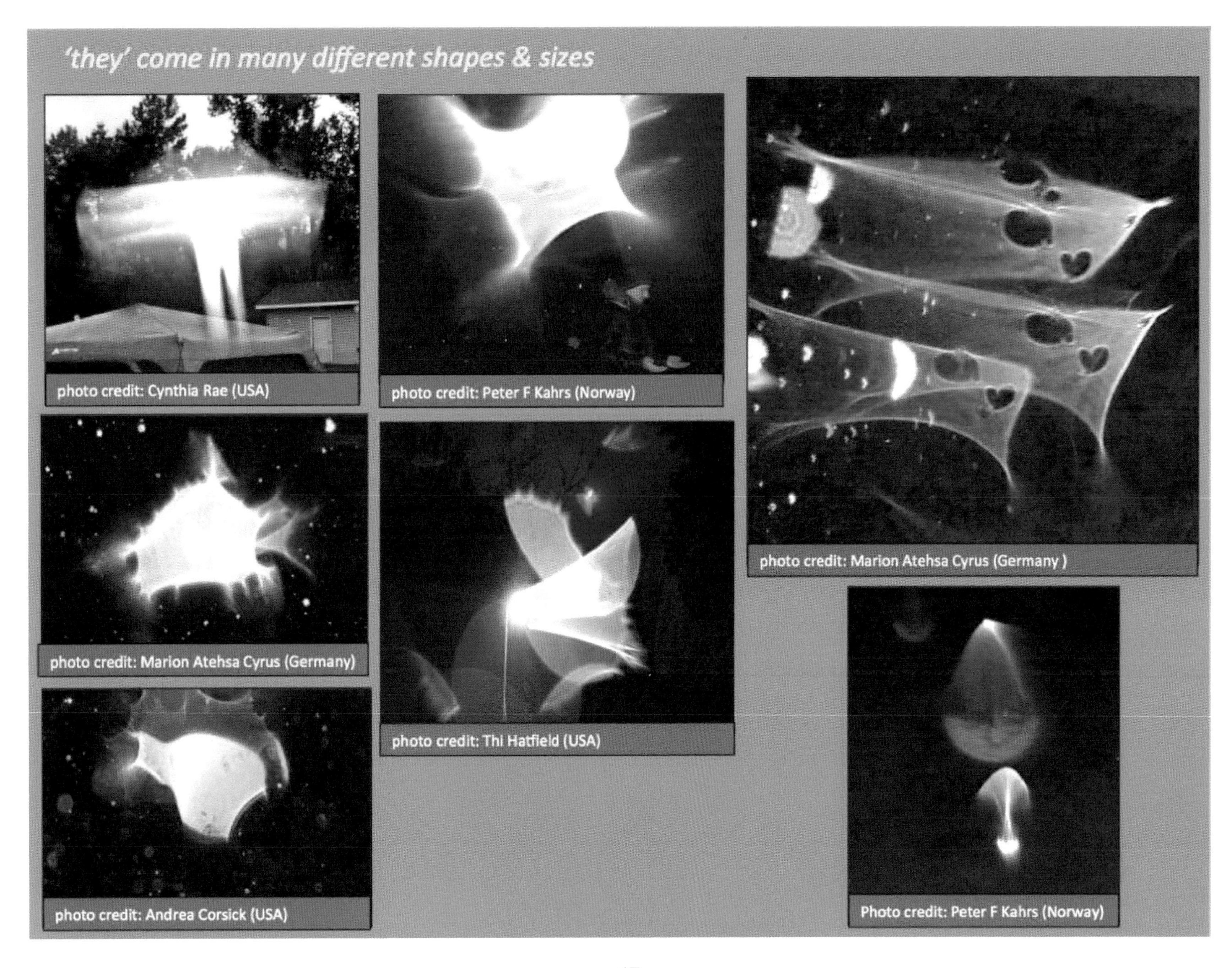
'they' come in many different shapes & sizes
photo credit: Cynthia Rae (USA)
photo credit: Peter F Kahrs (Norway)
photo credit: Marion Atehsa Cyrus (Germany)
photo credit: Marion Atehsa Cyrus (Germany)
photo credit: Thi Hatfield (USA)
photo credit: Andrea Corsick (USA)
Photo credit: Peter F Kahrs (Norway)

ORBS ARE TRAVEL CAPSULES…'TRAVEL CHAMBERS'

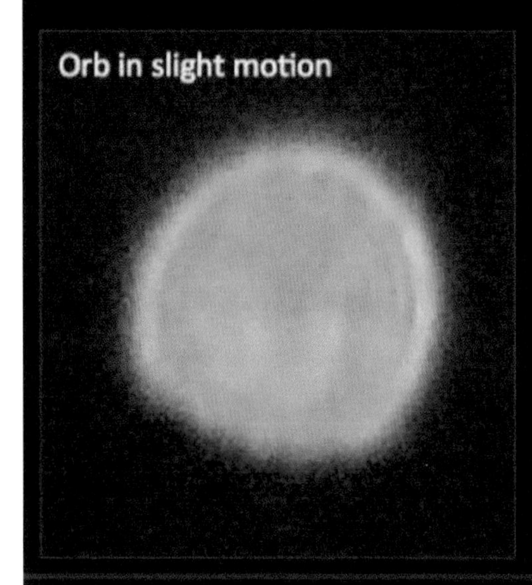

NOTE:

If we humans, could grasp the concept that, 'Orbs are travel Capsules or travel Chambers' from other frequencies, then it will be easier for us to understand that they are similar in many ways to our 3D Rockets & Spaceships that travel in 'our reality'.

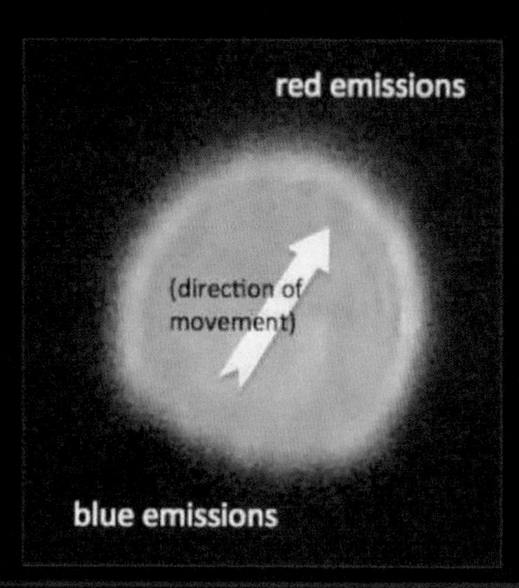

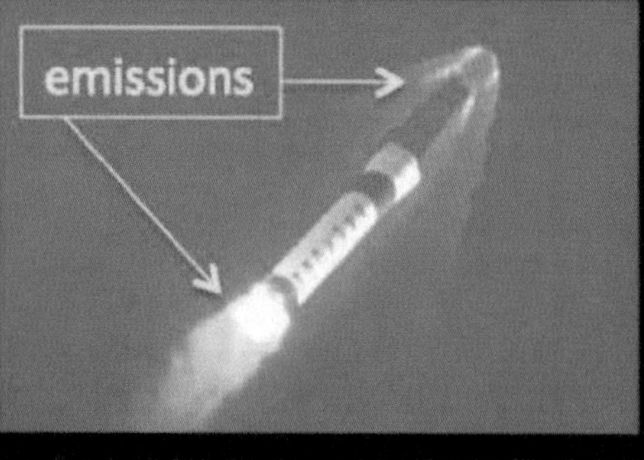

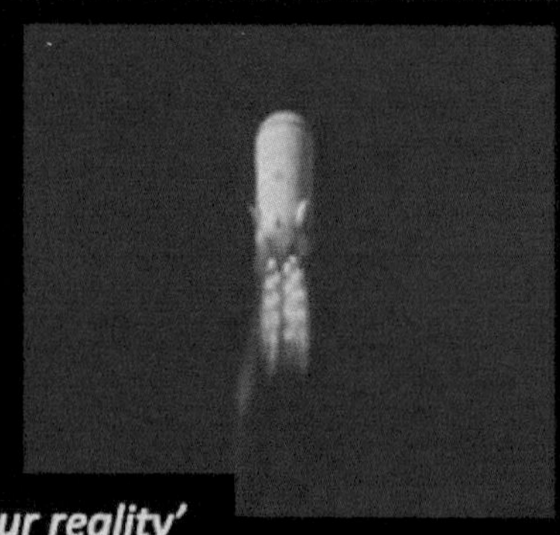

As a comparison: these are space rockets in 'our reality'

"THE DIRECTION OF MOVEMENT" IN ORBS

If an orb is in a 'complete stand-still position' (**Zero Motion**) ...it is usually perfectly round & has NO RED coloured external auras or emissions....this means that it has been at a stand-still for a while. (When an orb is at a stand-still, it is generally surrounded by neutral 'cooler colours').

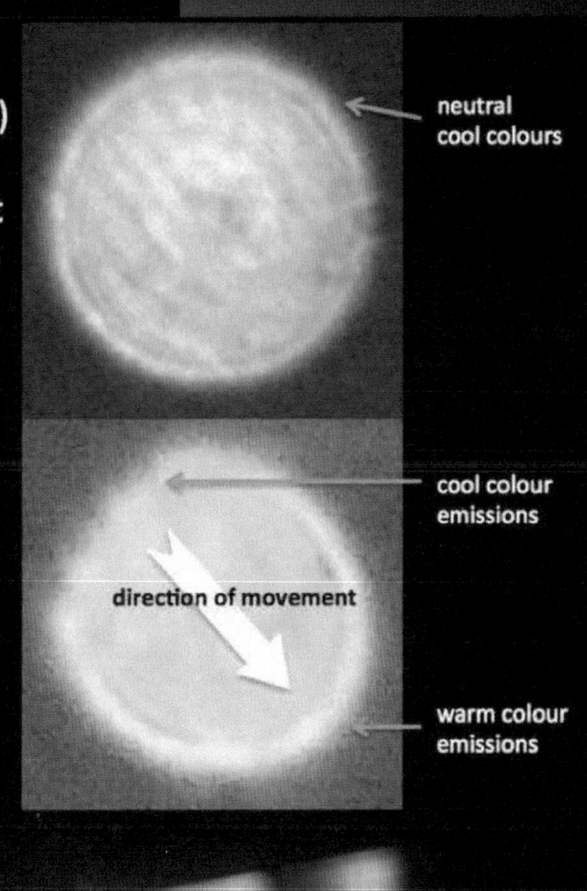

If an orb is in 'slight motion' (**Slow Hovering**)...
it is usually asymmetrical & has red & blue coloured external aura emissions...this means that it is travelling in the direction of the 'warm' colours (reds, oranges etc.) as it is moving away from the direction of the 'cool' colours (blues, violets etc.)

If an orb is in 'super motion' (**Extreme Velocity**)...
It is usually seen as a 'beam of light'...this means that it is travelling at 'unbelievable speeds'.

beams of light

MFT METHOD

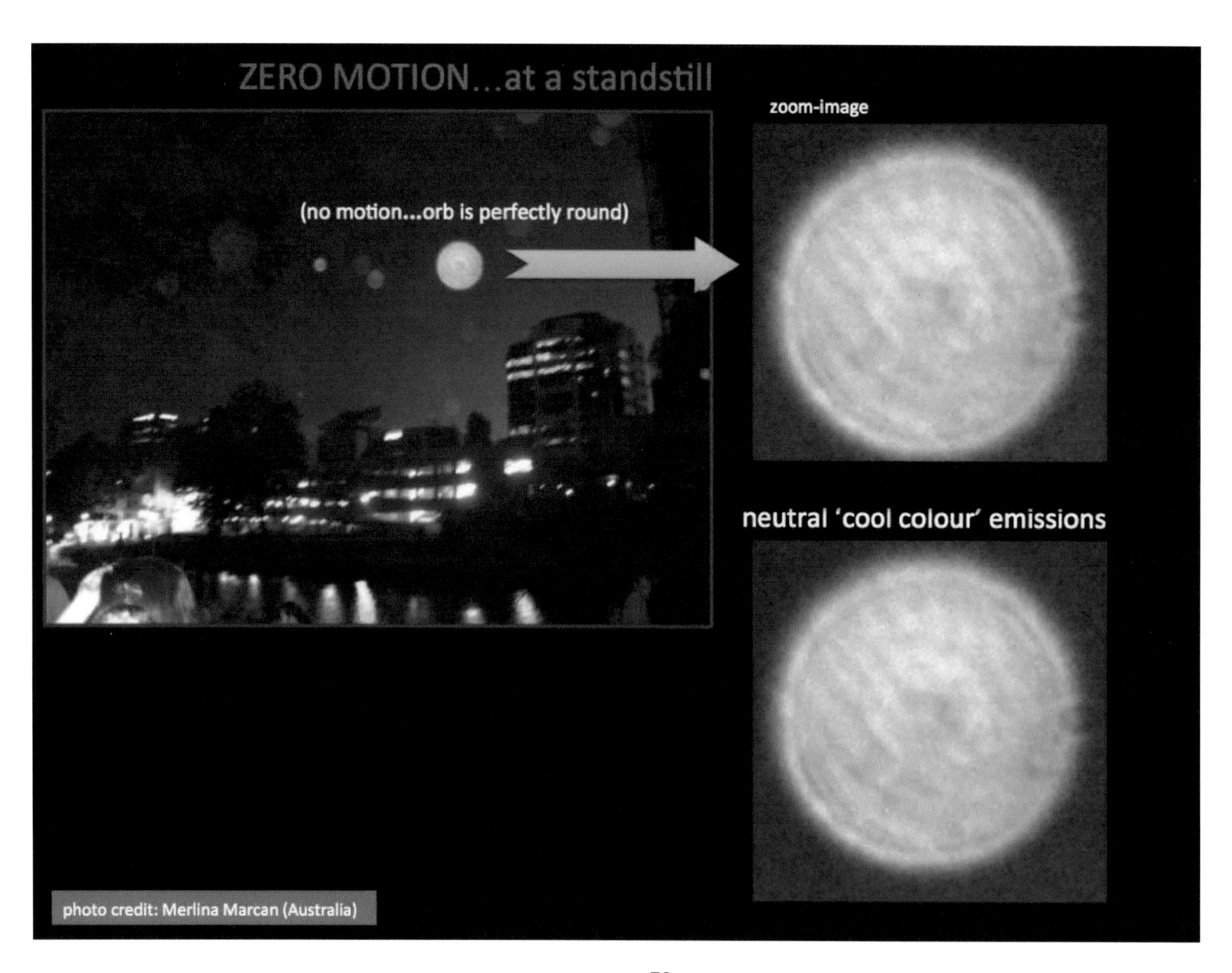
ZERO MOTION...at a standstill
zoom-image
(no motion...orb is perfectly round)
neutral 'cool colour' emissions
photo credit: Merlina Marcan (Australia)

Here are more examples of *"ZERO MOTION ORBS"* that I have captured:

These orbs have 'completely stopped' their motion and are in a stand-still mode. They are perfectly round and there are

NO WARM COLOUR EMISSIONS

(reds, orange etc.) on their outside capsules.

Emission colours around them are Neutral Colours, which mean that the orb has been at a stand-still for a while.

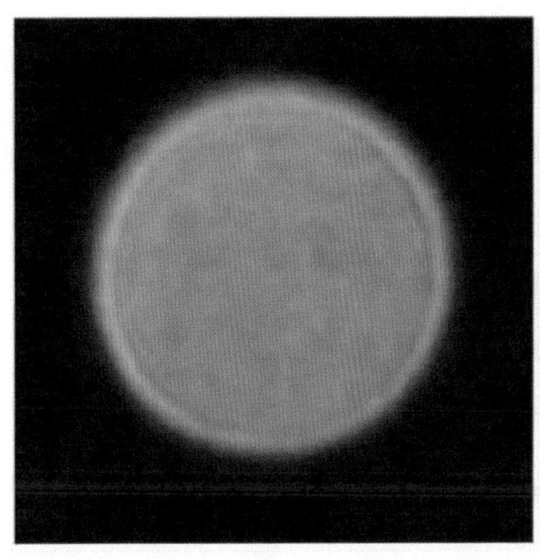

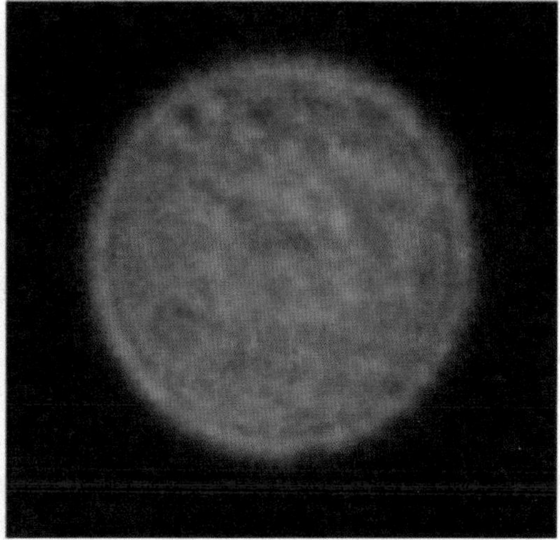

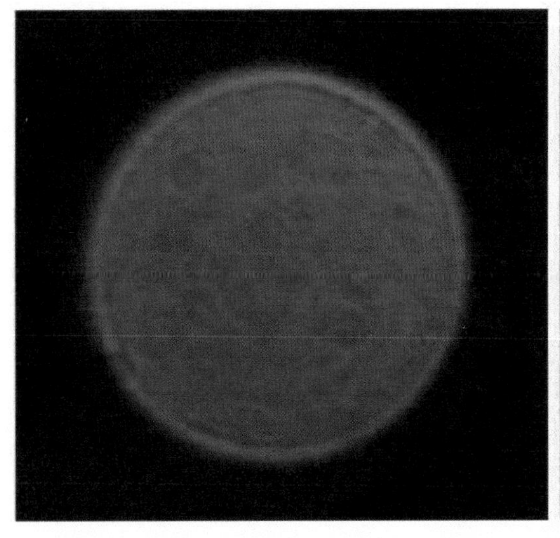

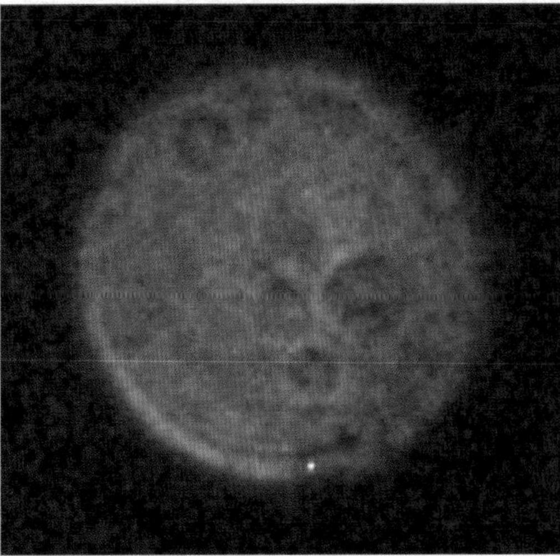

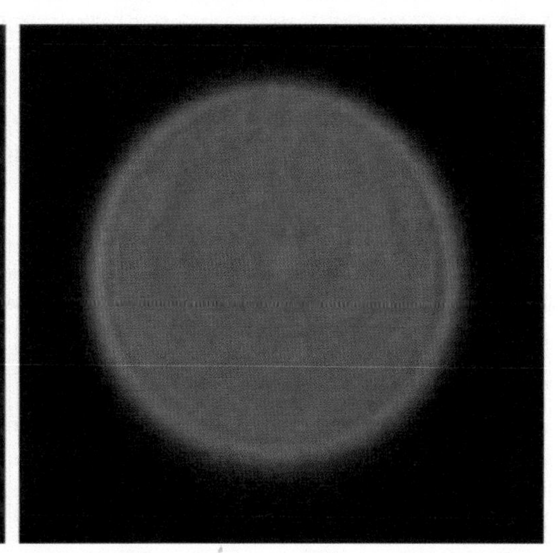

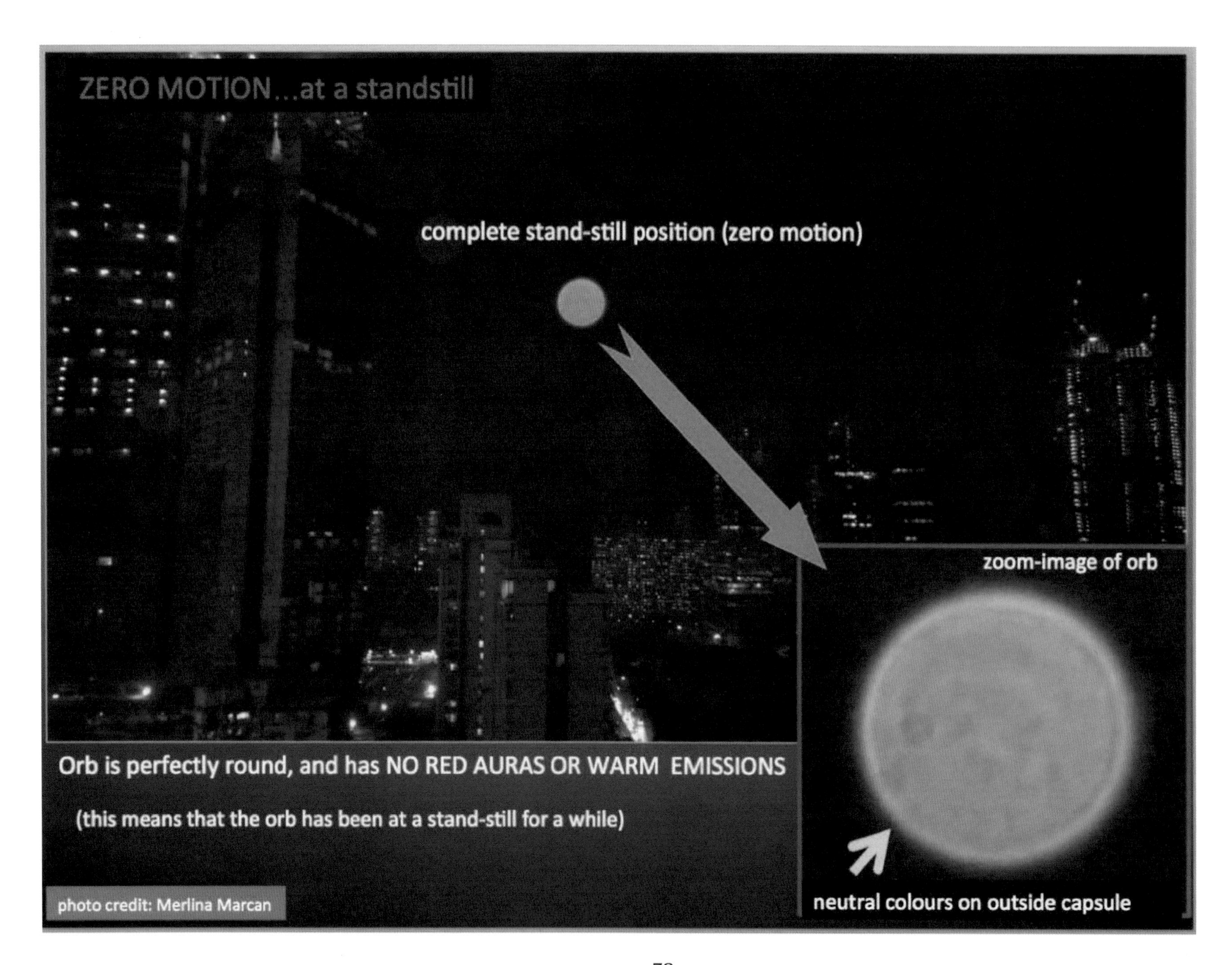
ZERO MOTION…at a standstill
complete stand-still position (zero motion)
zoom-image of orb
Orb is perfectly round, and has NO RED AURAS OR WARM EMISSIONS
(this means that the orb has been at a stand-still for a while)
photo credit: Merlina Marcan
neutral colours on outside capsule

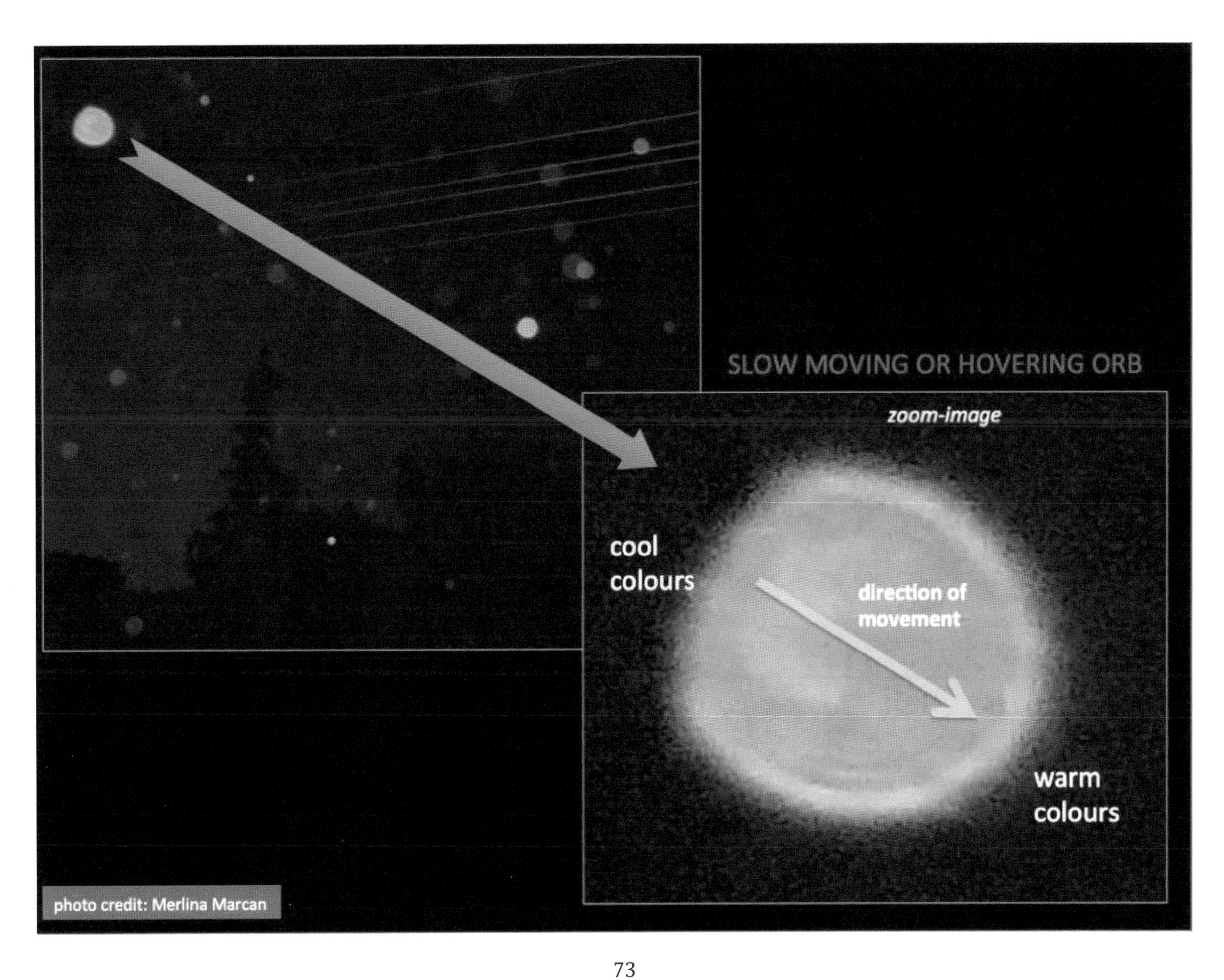
SLOW MOVING OR HOVERING ORB
zoom-image
cool colours
direction of movement
warm colours
photo credit: Merlina Marcan

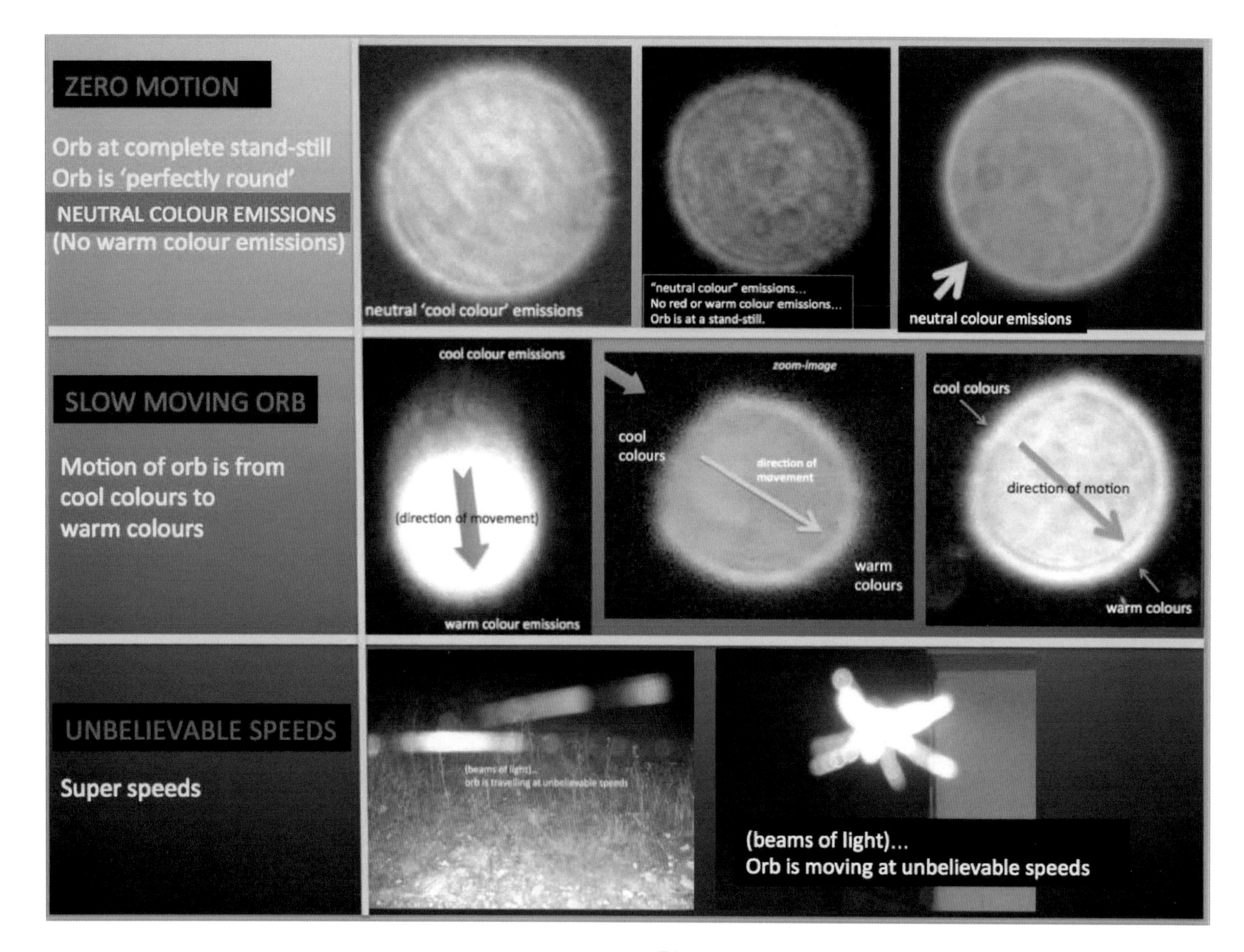
ZERO MOTION
Orb at complete stand-still
Orb is 'perfectly round'
NEUTRAL COLOUR EMISSIONS
(No warm colour emissions)
neutral 'cool colour' emissions
"neutral colour" emissions...
No red or warm colour emissions...
Orb is at a stand-still.
neutral colour emissions
SLOW MOVING ORB
Motion of orb is from
cool colours to
warm colours
cool colour emissions
(direction of movement)
warm colour emissions
zoom-image
cool
colours
direction of
movement
warm
colours
cool colours
direction of motion
warm colours
UNBELIEVABLE SPEEDS
Super speeds
(beams of light)...
Orb is moving at unbelievable speeds

Part 8
Categorization of Orbs

Orbs are *Cosmic Capsules*, or *Cosmic Vehicles* made of various light energies that may be unknown to us. Orb capsules can be like vessels, chambers or enclosures of extremely advanced 'auric protective shields'. These auric shields fortify and protect all the activity and information data that can be found inside the orbs. In other words, an orb is a *Protective Capsule Structure* that can change size, shape, colour, speed and direction depending on the mission that has to be accomplished. An orb can also phase in and out of our sight, change its atomic structure by switching its atoms and photons around to take on different forms, and it can easily transform from a light form to a solid 3-D structural form as it travels, simply by changing its frequency at will. ***In simple terms, an orb is a cosmic vehicle (rocket) made of light energy materials, thousands of years more advanced than our technology.***

I believe 'Cosmic Orb Vehicles' have specific '*rank levels*'. If we can understand the 'hierarchy levels' of these cosmic vehicles, then we may be able to understand *what 'Rank' of Sentient Beings can use that particular vehicle.* This should not be so surprising, because *we humans* also have 'vehicle hierarchy levels' too! And as an example, here is a simple question for you to answer…

Question:

"In order to protect a High Level VIP person, say for example, the President or Prime Minister of a country, a Royal Family Member, or a luminary of religion such as a Pope, what type of vehicle would you say that they would travel in"??
Would it be a run down old scrap car?? Or… would it be a highly fortified, bulletproofed, high security top quality vehicle??
I think we all know the correct answer to this question!
Therefore, if we were to see a highly fortified bulletproofed, top security vehicle coming down the street, we know it would most probably be a *High Ranking* person inside that vehicle that was very well protected. And orbs I believe, work on the same principles, where the 'thicker the outer auric membrane is', then the greater fortification the orb has, for its protection.

Through vision and observation, and after studying literally hundreds and hundreds of orbs, I have been able to see a 'definite pattern' and distinguish 'general characteristics' inside the orb phenomena. I have been able to categorize and divide them into 3 main groups, where each one of these groups, has specific trends, specific characteristics and specific patterns.

In some scientific fields and scientific research, ***Science is mainly based on 'predictions'.*** Scientists love to be able to predict the outcome of the tests, and the outcome of the results, so therefore a wide area of science is often based on the 'prediction results' obtained.
After nine years of my own personal research, I can now predict what type of information can be found in certain individual categories of orbs.

It was a challenge for me to find a name for the categorization of the Orb Phenomena, because I wanted a name that was unique, original, and very special, for these very special 'Multi-Dimensional Cosmic Structures' that we call *orbs*.
I am officially naming these three orb categories:

- ***Transcendent Module 1 …(TrM1)***
- ***Transcendent Module 2 …(TrM2)***
- ***Transcendent Module 3 …(TrM3)***

I chose to call the 3 categories of orbs by these names because '***TRANSCENDENT***' refers to various things that lie beyond the practical experience of ordinary people, and therefore cannot be easily understood or easily conceived by ordinary reasoning.
Transcendent may be something that is *beyond* our experience, but not necessarily beyond our potential of knowledge or the potential of further knowledge in the future.

'***MODULE***' refers to a self-contained object (as in this case the orb capsule), that can be combined with other compatible modules to form or to assemble a wider range of varied shapes and structures, it is like parts of a spacecraft that can operate independently of other parts when it is separated from them.
Hence the names, *Transcendent Modules 1,2 and 3*!
(See photos, PAGES 77-81)

Here are some of the main differences in three of the *'most common types'* of orbs

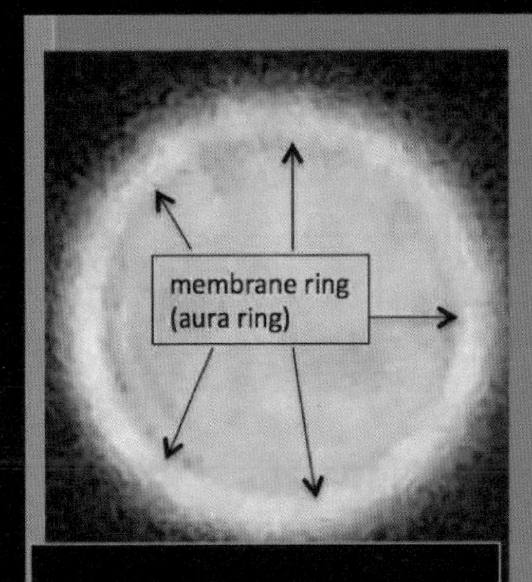

Transcendent Module 1
(TrM 1)

(TrM 1) orbs have a thick membrane (aura-ring) surrounding the orb.
This outer ring acts as a 'protective chamber', & I believe that only beings from the HIGHER REALMS travel within this category.
This category has 'very high luminosity of light' & they stand out from the rest.

- Sentient inside always have their bodies covered by the **'thick white protective cloudy-substance'.**
- Only multi-coloured heads & faces can be seen, & are sparsely populated.

Transcendent Module 2
(TrM 2)

(Trm 2) orbs have a very thin outer membrane-ring, & may sometimes be very hard to see. I believe that only beings from the MID-ASTRAL REALMS travel within this category.
This category is not very luminous, but it is still fairly noticeable.

- Sentient inside do not have much of the 'thick white protective cloudy substance'.
- Smaller sentient heads can be seen and are increasing in numbers.
- Often wearing headgear of their epoch or era.

Transcendent Module 3
(TrM 3)

(TrM 3) orbs have absolutely **NO protective membrane ring** or membrane layers.
This category is very hard to spot in photos, because there is NO light reflection.
These orbs are like *'dark patches'* in the sky, and I believe that only beings from the LOW-ASTRAL REALMS travel within this category.

- Sentient inside this category are extremely small in size & are very over-crowded.
- Literally, like a 'can of sardines'.

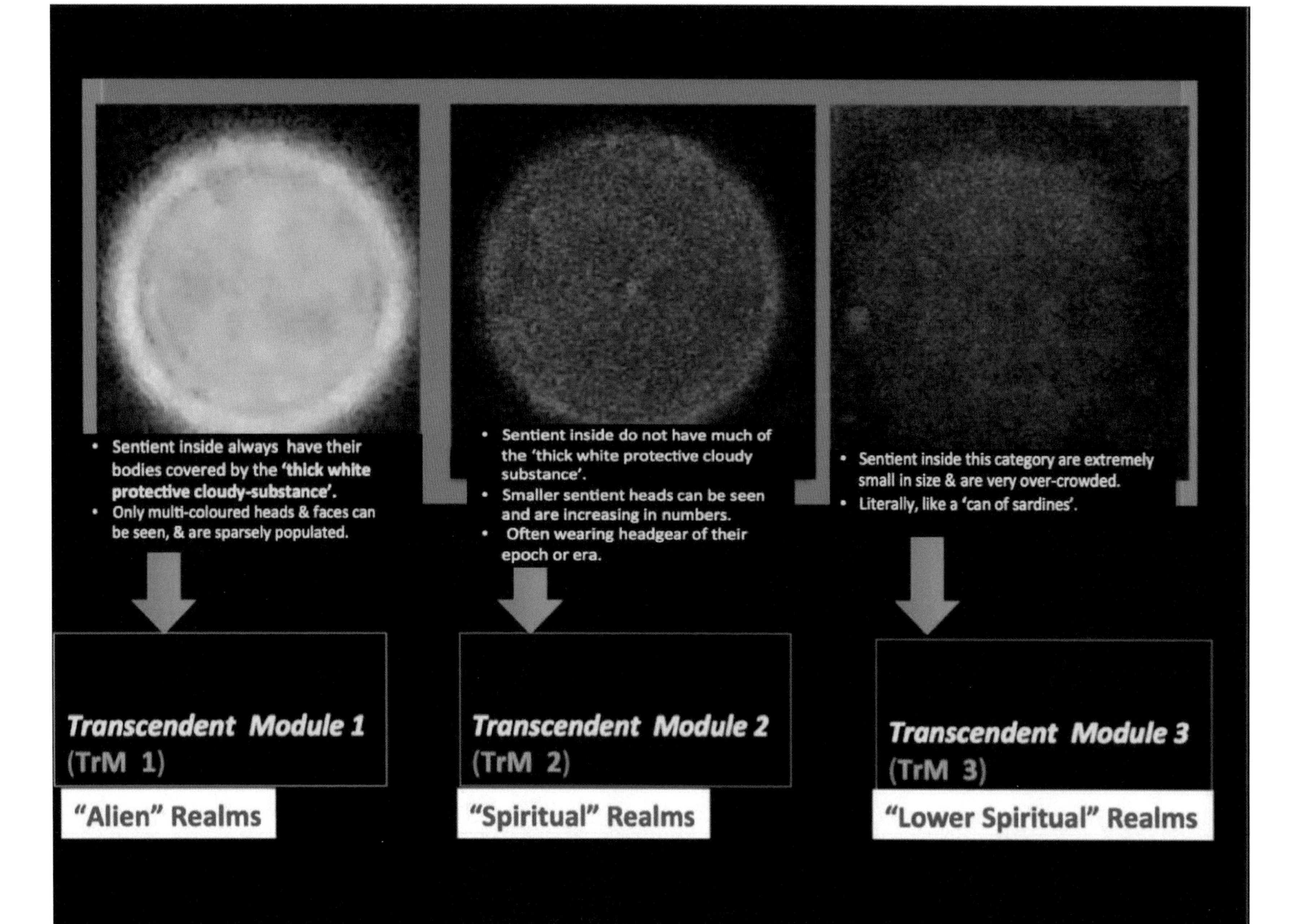
• Sentient inside always have their bodies covered by the 'thick white protective cloudy-substance'.
• Only multi-coloured heads & faces can be seen, & are sparsely populated.
• Sentient inside do not have much of the 'thick white protective cloudy substance'.
• Smaller sentient heads can be seen and are increasing in numbers.
• Often wearing headgear of their epoch or era.
• Sentient inside this category are extremely small in size & are very over-crowded.
• Literally, like a 'can of sardines'.
Transcendent Module 1
(TrM 1)
"Alien" Realms
Transcendent Module 2
(TrM 2)
"Spiritual" Realms
Transcendent Module 3
(TrM 3)
"Lower Spiritual" Realms

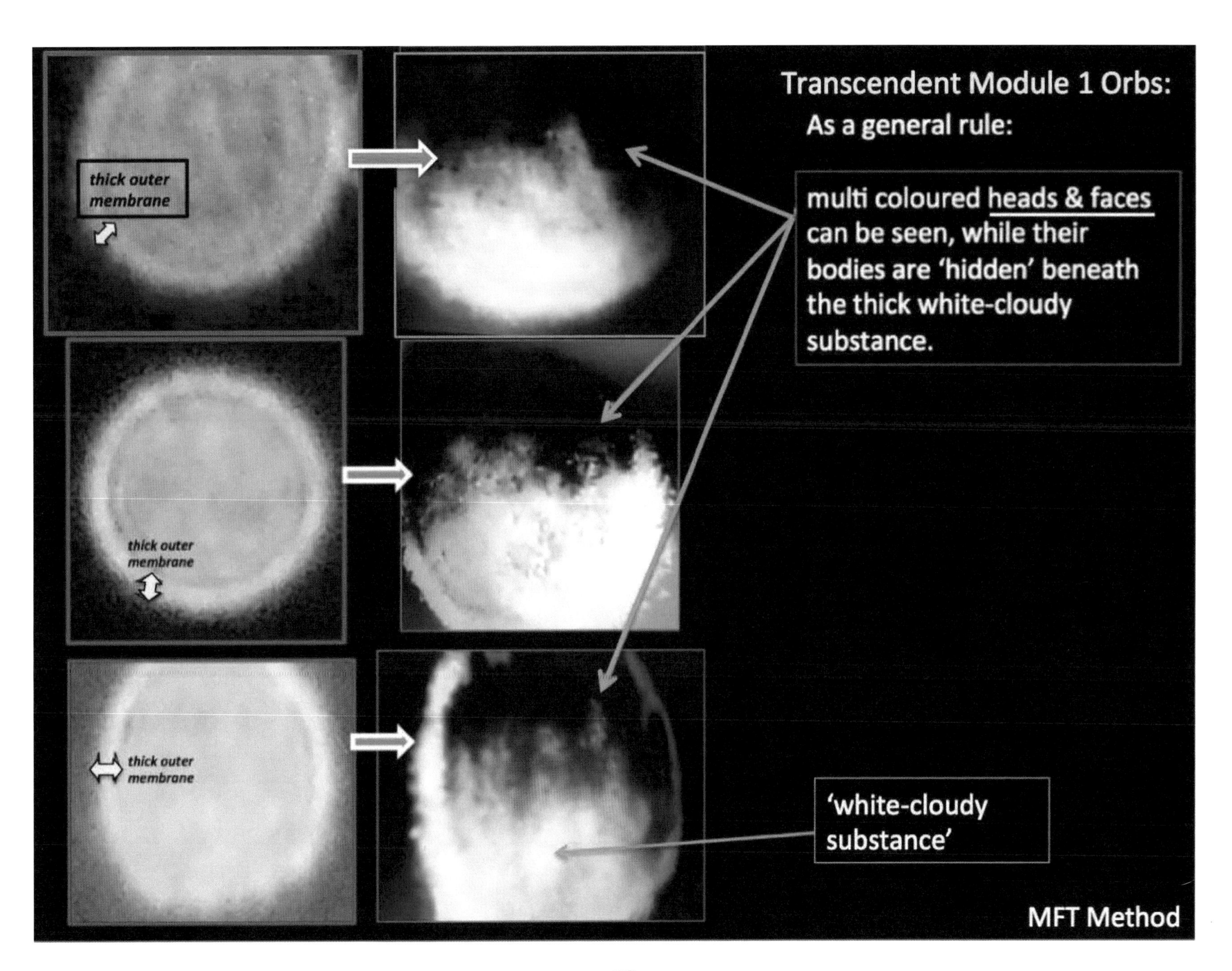
Transcendent Module 1 Orbs:
As a general rule:
multi coloured heads & faces can be seen, while their bodies are 'hidden' beneath the thick white-cloudy substance.
thick outer membrane
thick outer membrane
thick outer membrane
'white-cloudy substance'
MFT Method

Examples of *TrM1 orbs:* *"Transcendent Module 1 orbs"* (Alien Realms)

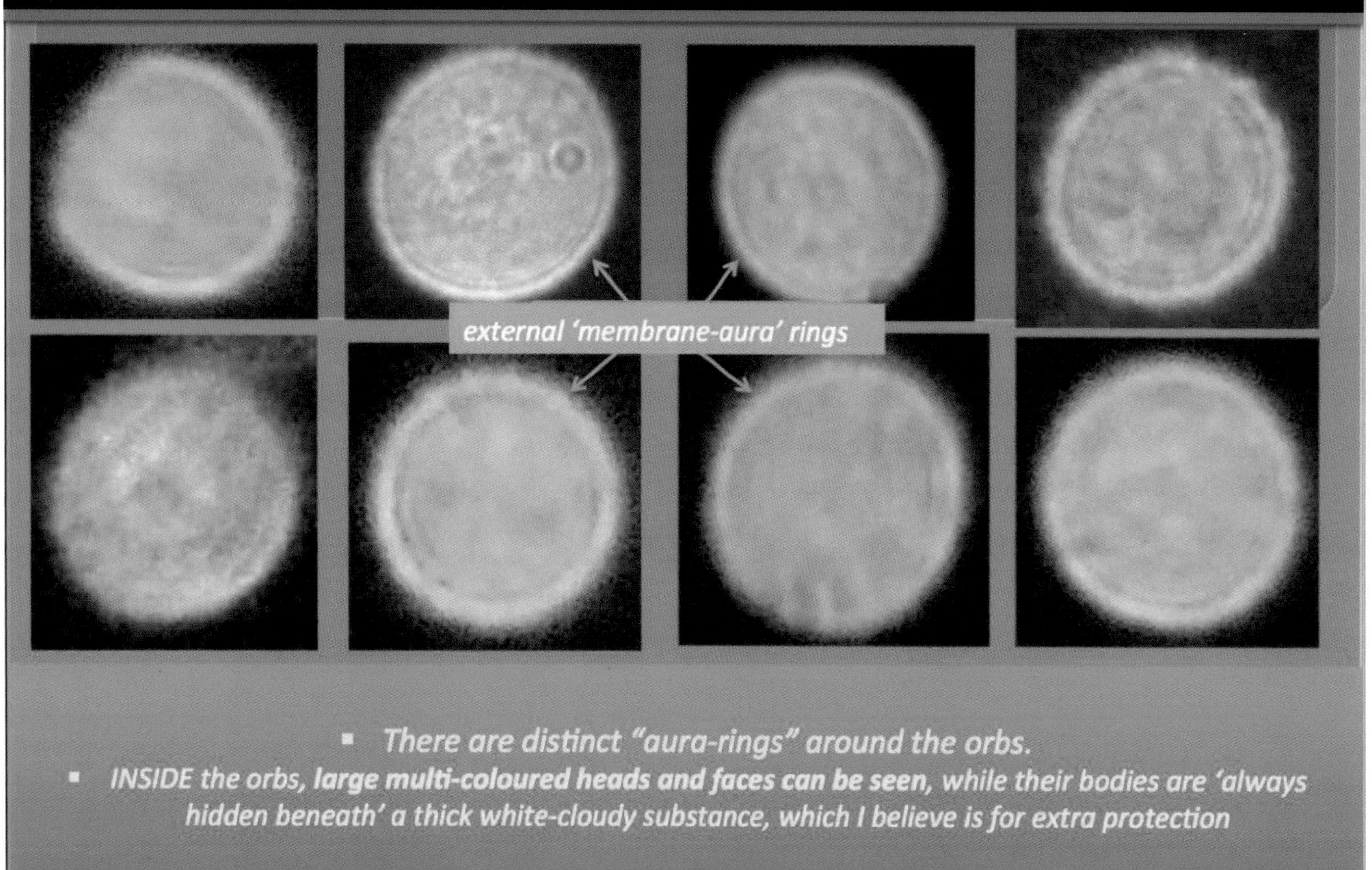

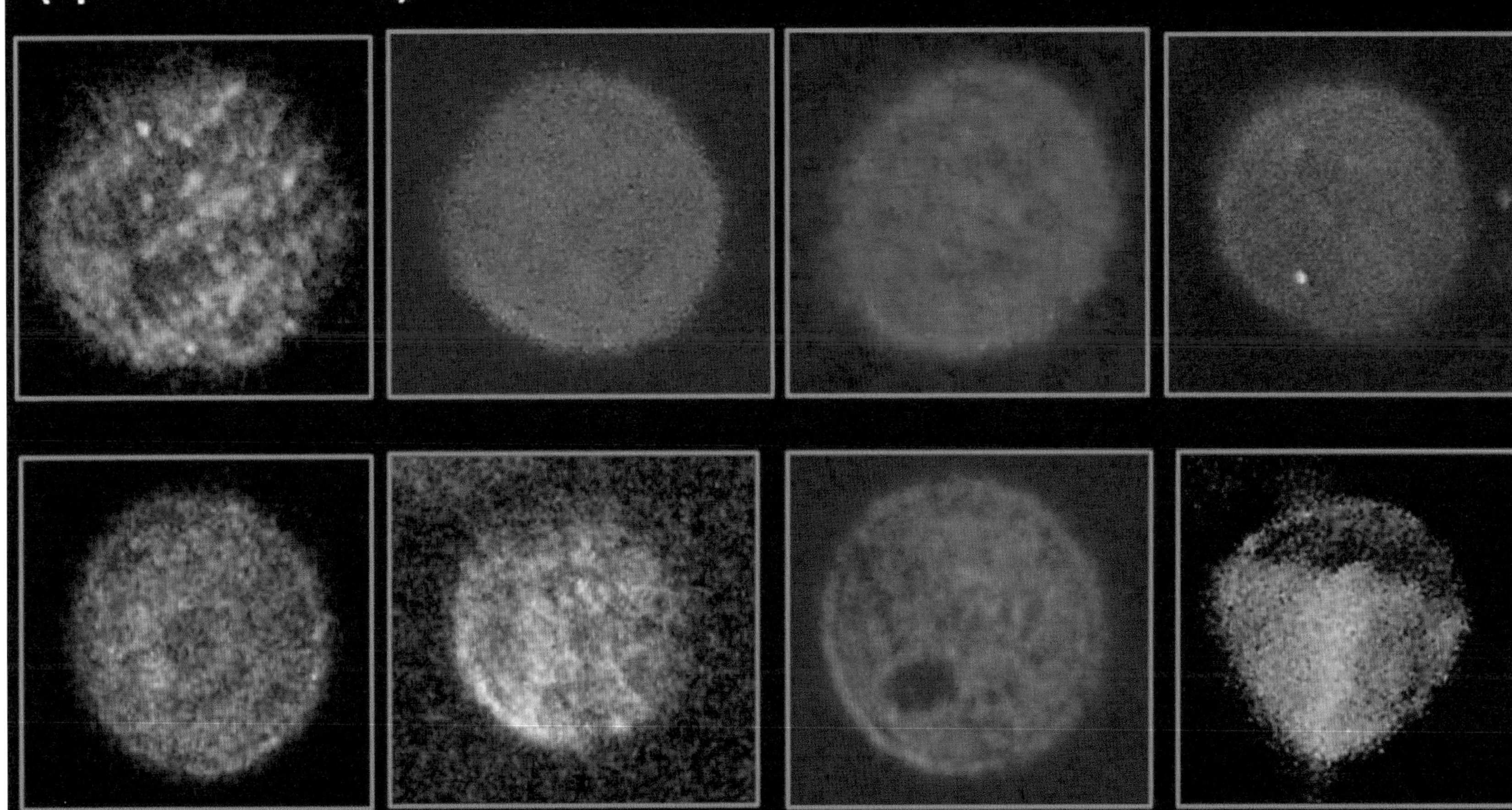
Examples of TrM2 orbs:
(Spiritual Realms)
"Transcendent Module 2 orbs"
This category is not very bright or luminous, but is still fairly noticeable.
Inside the orbs, smaller multi-coloured heads and faces can be seen, often in 'epoch type' clothing or headgear.
The 'thick white-cloudy substance' is diminishing.

Part 9
Pareidolia & Matrixing

I will show you visual evidence that 'The Orb Phenomena' carries a vast and astronomical array of data and information that can be seen inside these cosmic crafts. My Pioneering Research uses methods, ideas and hypotheses that have never been ventured before and as a *citizen pioneer* in this field; I am instigating and introducing the internal and intramural foundations of orbs.
But before I start on this pioneering venture, I would like to talk about an important factor, which can sometimes arise in pioneering research. This important factor is called *pareidolia or matrixing.*

'Pareidolia' is a vague image that causes a psychological phenomenon of seeing images of faces, animals, or objects that are not there. It is a psychological tendency for the observer to interpret abstract images of inanimate objects into familiar perceptions, such as seeing faces and animals for example, in cloud formations, the moon's surface, in trees and nature, and other various everyday objects etc.

BUT...**IN PIONEERING RESEARCH, 'PAREIDOLIA' CAN SOMETIMES BE A VERY FINE LINE BETWEEN 'REALITY' AND 'IMAGINATION'.**
As with most pioneering discoveries, quite often the prototype images, the prototype photos, the prototype scans and the prototype experimental data are often very low quality, often grainy or blurry images, which may sometimes need a little imagination to make sense of the images being researched. We have often seen this in the pioneering study of Astronomy, Astrophysics, Ultrasound Imaging, and Microscopy Imaging to mention a few. As quite often in pioneering research, the optical textures and optical granularity of the images need years, and in some cases even decades before the image improvements can be seen.

Here are a few examples where Pioneering Prototype Images show the *fine line between 'reality and imagination'* that have *gradually evolved* and have made headway, with years and even decades of time, by developing and improving the images with more research, more research funding of millions of dollars, and eventually more advanced technology. This is *science in the making*!
(See photos, PAGES 83-84)

Photo credits: NASA

The first image of ***Earth*** from space
Explorer 6, 1959

Explorer 6 And The First Satellite Picture Of Earth in 1959...

Pioneering images

An image of ***Earth*** in 1972, taken by the crew of the Apollo 17 spacecraft.

In Pioneering Research, developments do not often happen overnight. In this case of space science for example, it took millions of dollars for better technology and better equipment, over a span of many decades, with the world's greatest scientists and greatest high-calibre minds working together, to eventually get images such as this 'Blue Marble' photo of Earth.

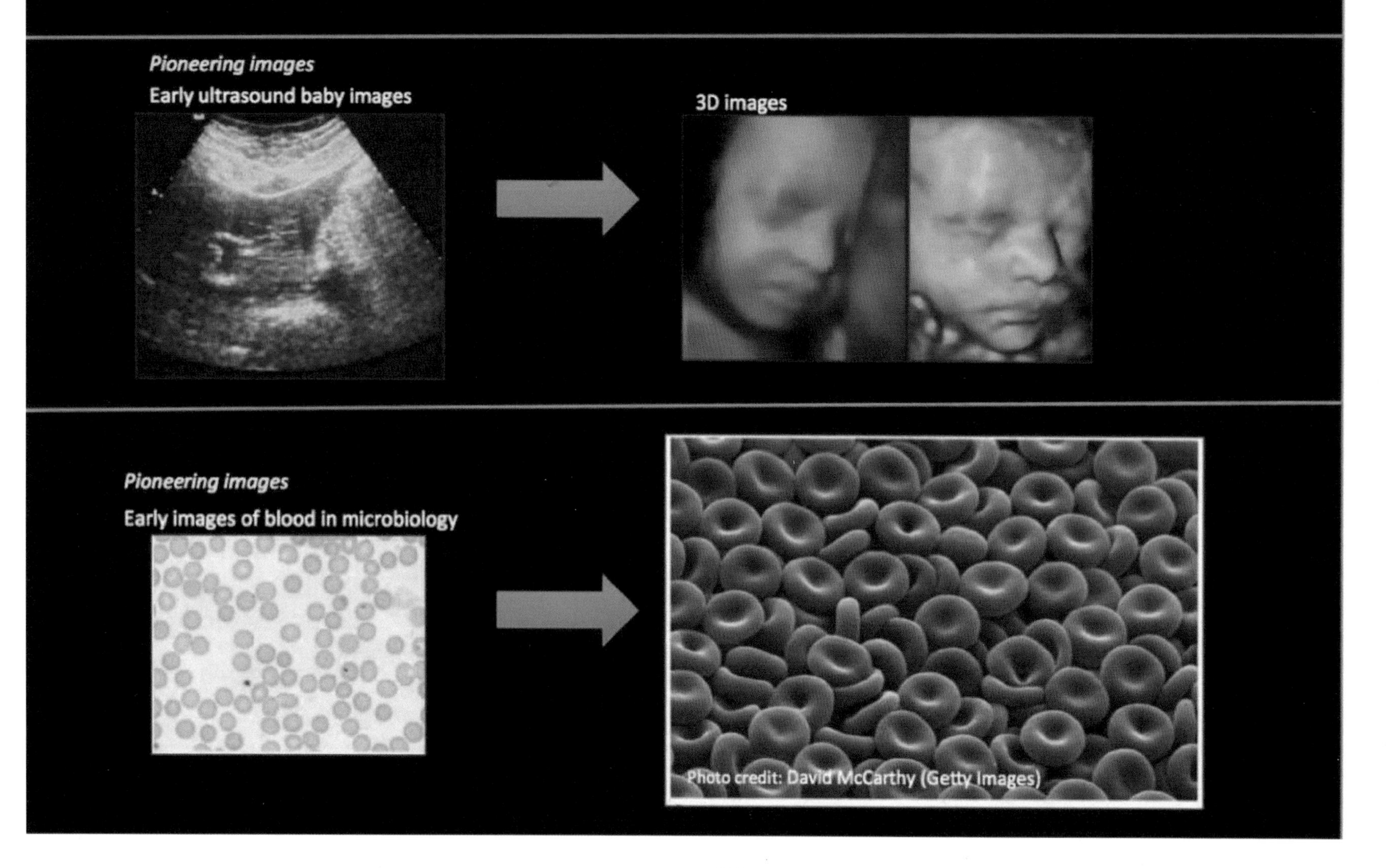
Ultrasound and Microbiology over the last few decades, have undergone extensive developments.
In Pioneering Research, developments do not often happen overnight. In this case of ultrasound and microbiology science, it took millions of dollars for better technology and better equipment, over a span of many decades, with the world's greatest scientists and greatest high-calibre minds working together, to eventually get images such as these 3-D imaging.
Pioneering images
Early ultrasound baby images
3D images
Pioneering images
Early images of blood in microbiology
Photo credit: David McCarthy (Getty Images)

So now that I have given a little brief about 'pareidolia' and a little brief about 'pioneering research images', I think it is time to look at some of my own research images. The images you will see here are absolutely *not* pareidolia, and as far as pioneering research goes, the answer is *yes*, there is much potential for further extensive developments in the near future to get better images of data inside true orbs. My research work is still in its infancy stages, but once this infancy stage begins to grow and develop, the 'orb phenomena' may prove to be one of the most intriguing, enthralling and thought-provoking phenomena to be witnessed and understood by mankind. It will open up new pathways to understanding '*new horizons*'.

My work and the images that I am capturing are as *pioneering* as it can get. My images are authentic, with new and fresh hypotheses that may be challenging to many. There is a vast *Unexplored Universe* out there, it is overflowing with life and many different life forms, and since the early 20th century, there has been an ongoing search for signs of Extraterrestrial Life. Aliens, spiritual beings and inhabitants of other dimensions can go far beyond our current understanding of Laws of Physics. Beings from these different dimensions and realities can effortlessly glide through oceans of different light and atmospheric conditions to come visit us here on Earth.

The Universe is like a wireless network of data information that is being projected throughout the cosmos. We can compare this wireless network of the Universe to a simple transistor radio that can display various static sounds from various different frequencies. By simply fine-tuning and changing the dial to different wavelengths, we are able to have access to different radio stations. And the same applies to the Universe, when we tap into other dimensional realities, we can see images of beings from these other realities. I often refer to these beings as ***'Sentient Beings'*** because the term *sentient* is an intelligent life form with consciousness and will. As we will see, some of these sentient may resemble human-type forms, some may look semi-human, and some may altogether look strange and alien-looking, unlike anything that we can compare. We will see them in what will look like *Holographic Displays of cosmic sentient beings inside the orbs. 'They know'* when they are being photographed, and we will often see them looking straight into the camera, and 'posing' for us as they are being photographed.

Sentient Beings inside orbs are very 'inquisitive' and they enjoy interacting with us. From my experience and observations over the last ten years, the more we become attuned to these other dimensions, then the more responsive they become to us, as they can feel, sense, and understand our intentions. We must always remember, that when we are Orb Hunting for photographs, we should never try to provoke or intimidate them, but instead; we must show respect and patience and they will eventually appear in our photos. There is no need to be afraid of them. Whether they are aliens, extraterrestrial, inter-dimensional beings, hybrid races or spiritual celestial beings, ***they are not here to harm us.*** They are here to observe and interact with us.

Through simple observation of my research photos of the sentient beings, there are strict '*Hierarchy Levels*' inside orb crafts where we will often see ***discrimination in 'sizes' of the beings***. For example, in a same particular orb, we may see an *extremely large* dominant sentient face or head, and the rest may be *smaller in size*, or the smaller sentient may *overlap each other* so that they will not have as much space as the dominant sentient. Sometimes in orb crafts, we will see only a few large heads that are perhaps the higher ranking sentient, and sometimes we will see many numerous smaller heads that overlap each other that may have lower ranks.

Let us compare this for example, with many of the Ancient Egyptian wall murals that can be found in ancient temples and tombs. The Higher Rank was depicted as larger images, whereas the Lower Ranks were much smaller in size.

Part 10
Transcendent Module 1 orbs: (TrM1 orbs)

Module 1 orbs are the orb crafts, the orb vehicles or the orb vessels that have the brightest luminosity. They are extremely refulgent and bright and they stand out from the rest. Module 1 orbs are the ***'WOW factors in photos'***. If you capture a 'Module 1' in your photo, your initial reaction when seeing it will most likely be a "WOW! Look what I captured!" or something to this effect. So Module 1 Orbs are the WOW FACTORS! We will often see them in colours of brilliant whites, golden, silvery, luminous blues and many other standout colours that will capture our attention. On closer observation of Module 1 orbs, when we zoom into them, one of their very distinct characteristics is that they often will have a thick and distinct bright *'aura-ring' or 'thick membrane' around them.*

This outer-ring acts as a protective chamber that fortifies and encloses as a protective shield. I believe that the sentient activity and data information inside these modules are very well protected as they travel through different light and atmospheric pressures and cosmic conditions. I believe this category of orbs is the travel means by beings from the *Far Distant Realms that have a long way to travel*, and that they are from the Extraterrestrial Alien and Inter-Dimensional Realities from other galaxies and star-systems. This is why I believe they need that extra thick 'aura-shield-ring-membrane' around them for protection.

Sentient activity inside these modules, always have a very *'thick white-cloudy substance' that covers and protects them*. Large multi-coloured ***Heads and Faces*** will be seen, but never a torso or body, as it seems that body parts are non-existent, or non-relevant, or very rare. It is the 'Heads and Faces' that seem to be of great importance, and that they are well protected inside and within the boundaries of the surrounding aura membrane-ring of the orb. Sometimes, smaller size heads can be seen around the dominant large size head or heads (that are perhaps of a higher rank), as if they are surrounding and protecting them. I have found that many of the Module 1 beings, can often be seen as if they are '*smiling*', as they pose for the photo.

GENERAL CHARACTERISTICS FOR TrM1 Orbs:

- TrM1 orbs are the bright luminous WOW factors in photos.
- TrM1 orbs are often in colours of brilliant whites, golden, silvery blues and many other stand out colours.
- TrM1 orbs are often surrounded by a thick '*aura-membrane-ring*' around them for protection.
- Inside the orb, *large multi-coloured Heads and Faces* can be seen.
- No bodies can be seen as a '*thick white cloudy substance*' covers them.
- Module 1 orbs are *sparsely populated*, and there is never over-crowding as compared to the other categories.

(See photos, PAGES 88-95)

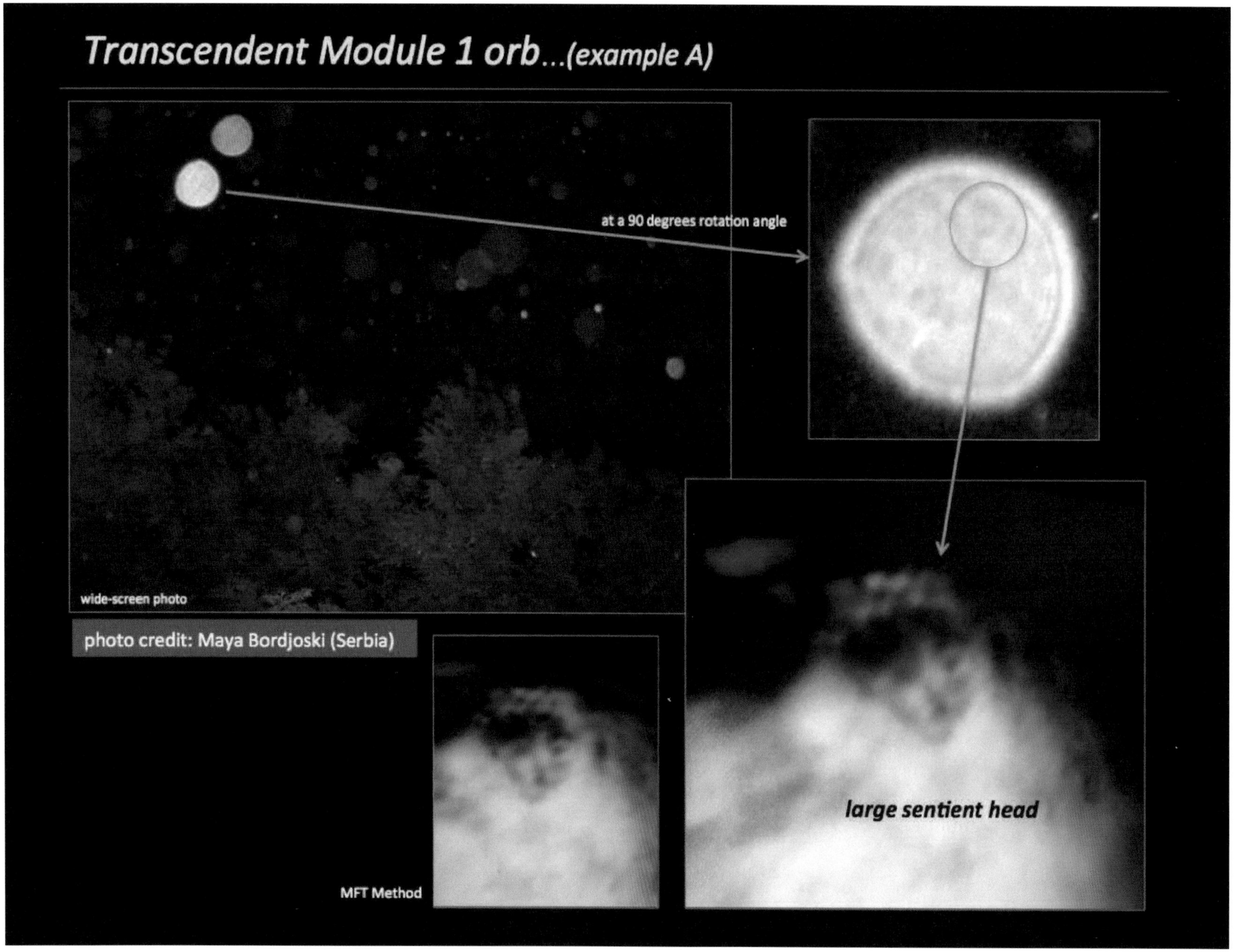

Transcendent Module 1 orb...(example A)

at a 90 degrees rotation angle

wide-screen photo

photo credit: Maya Bordjoski (Serbia)

MFT Method

large sentient head

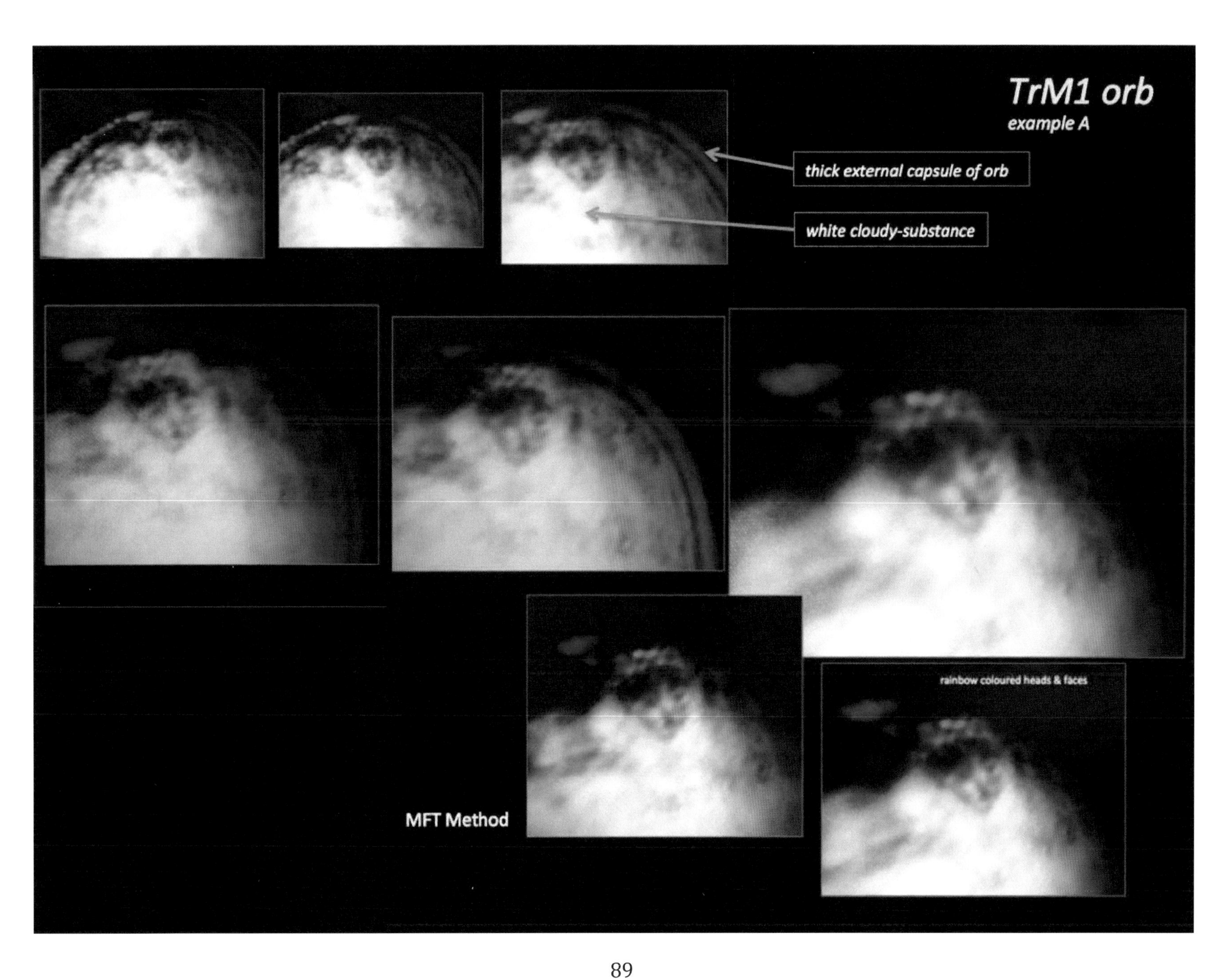
TrM1 orb
example A
thick external capsule of orb
white cloudy-substance
rainbow coloured heads & faces
MFT Method

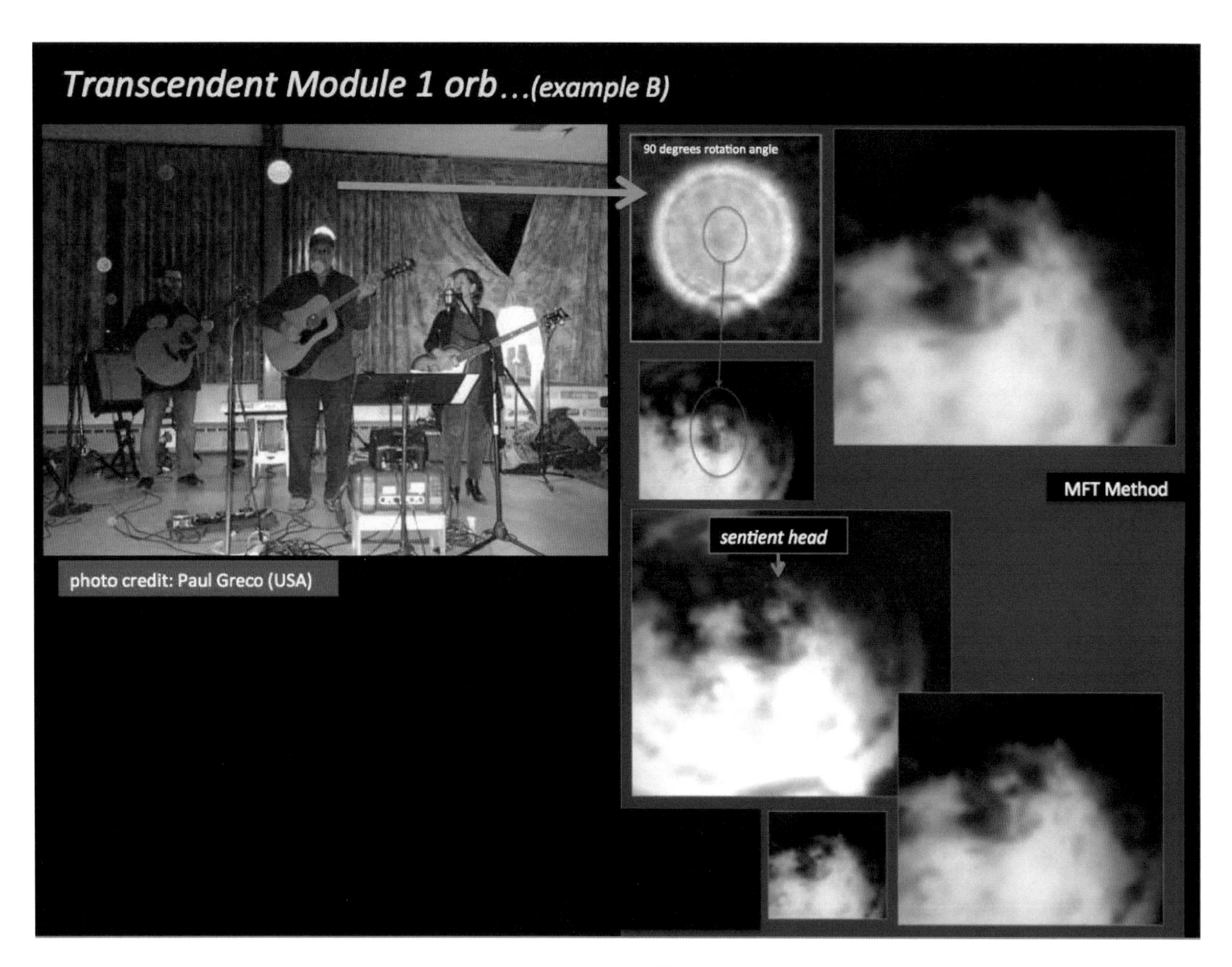
Transcendent Module 1 orb...(example B)
90 degrees rotation angle
MFT Method
sentient head
photo credit: Paul Greco (USA)

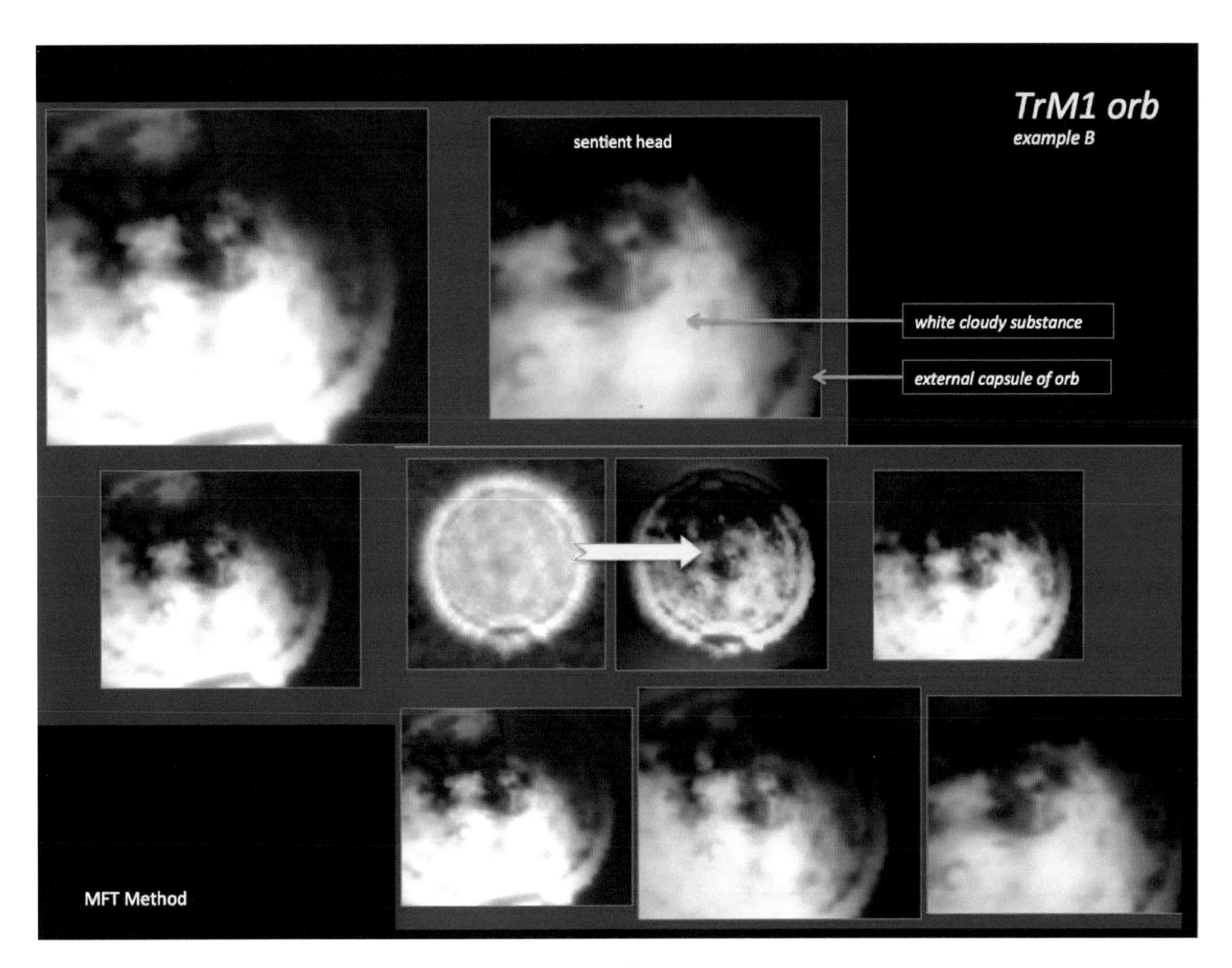
TrM1 orb
example B
sentient head
white cloudy substance
external capsule of orb
MFT Method

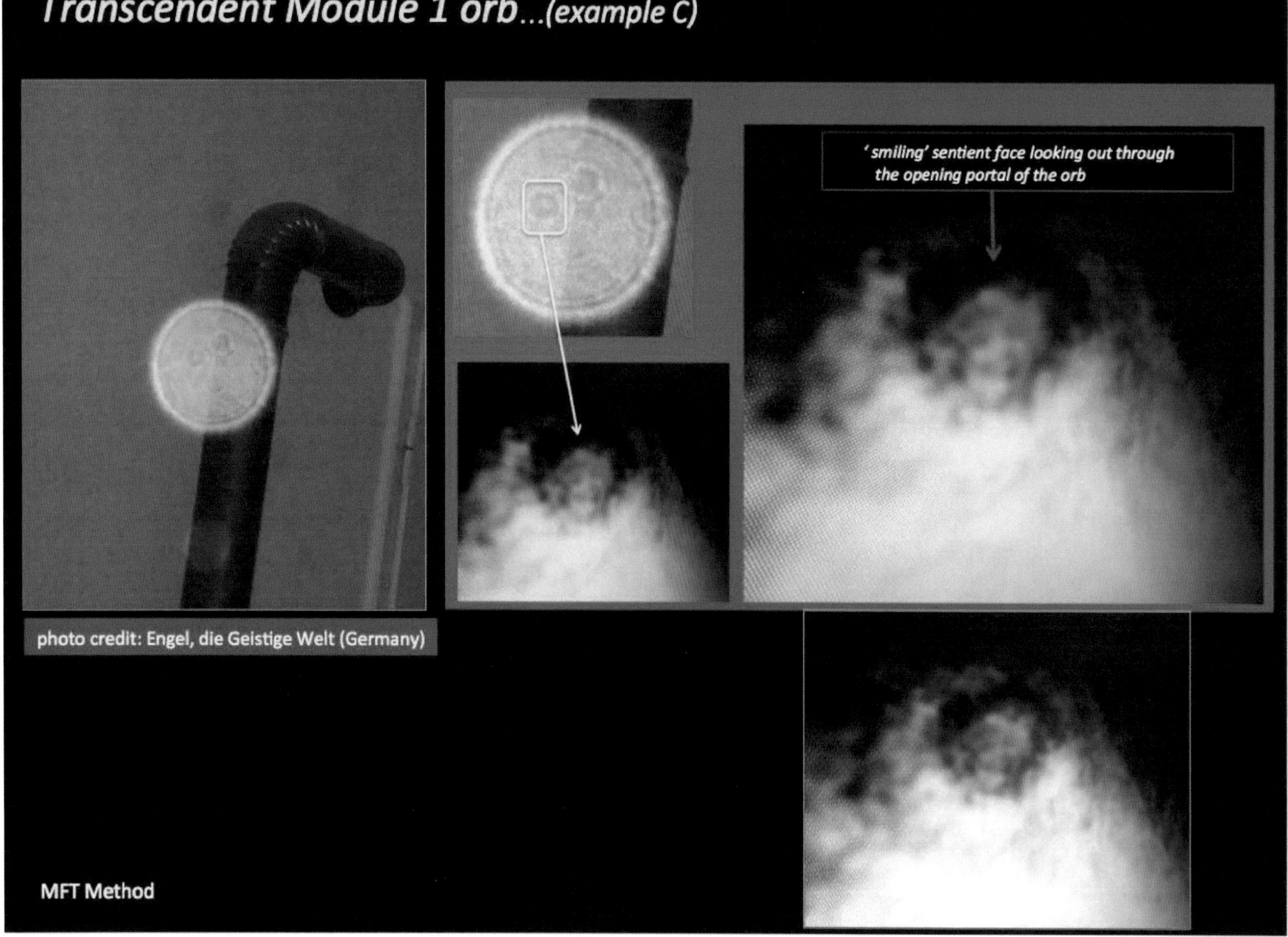
Transcendent Module 1 orb...(example C)
'smiling' sentient face looking out through the opening portal of the orb
photo credit: Engel, die Geistige Welt (Germany)
MFT Method

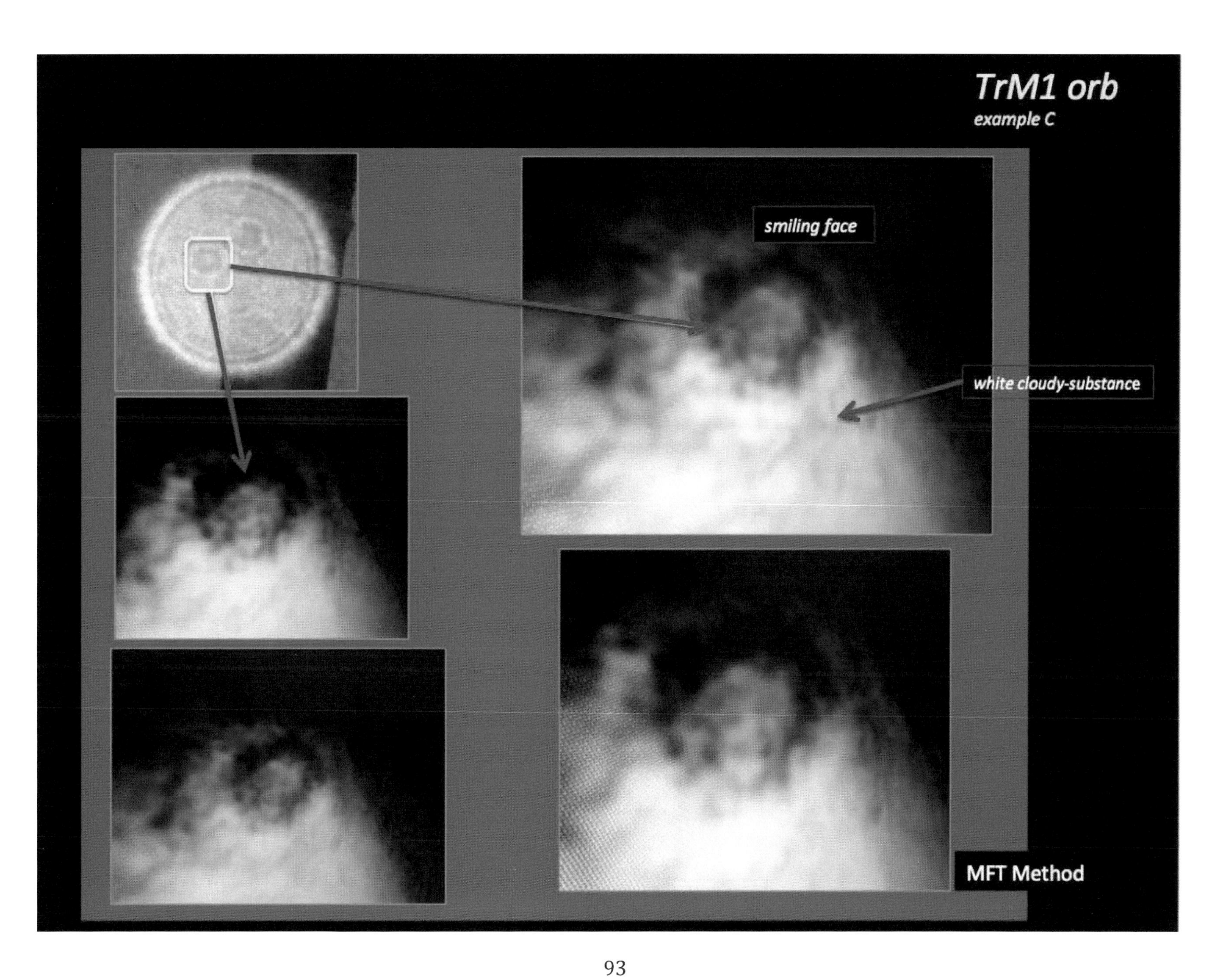
TrM1 orb
example C
smiling face
white cloudy-substance
MFT Method

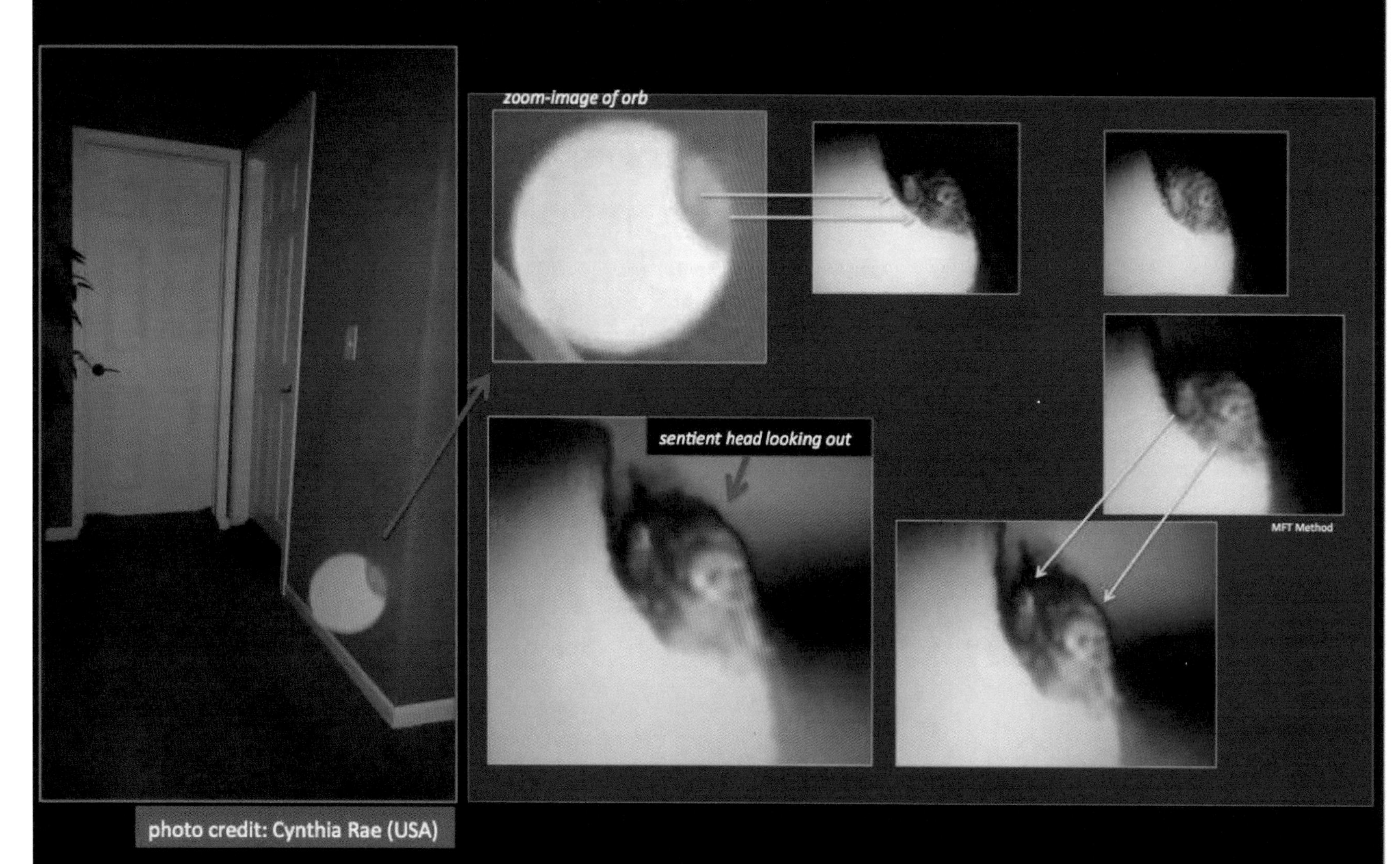
Transcendent Module 1 orb...(example D)
zoom-image of orb
sentient head looking out
MFT Method
photo credit: Cynthia Rae (USA)
MFT Method

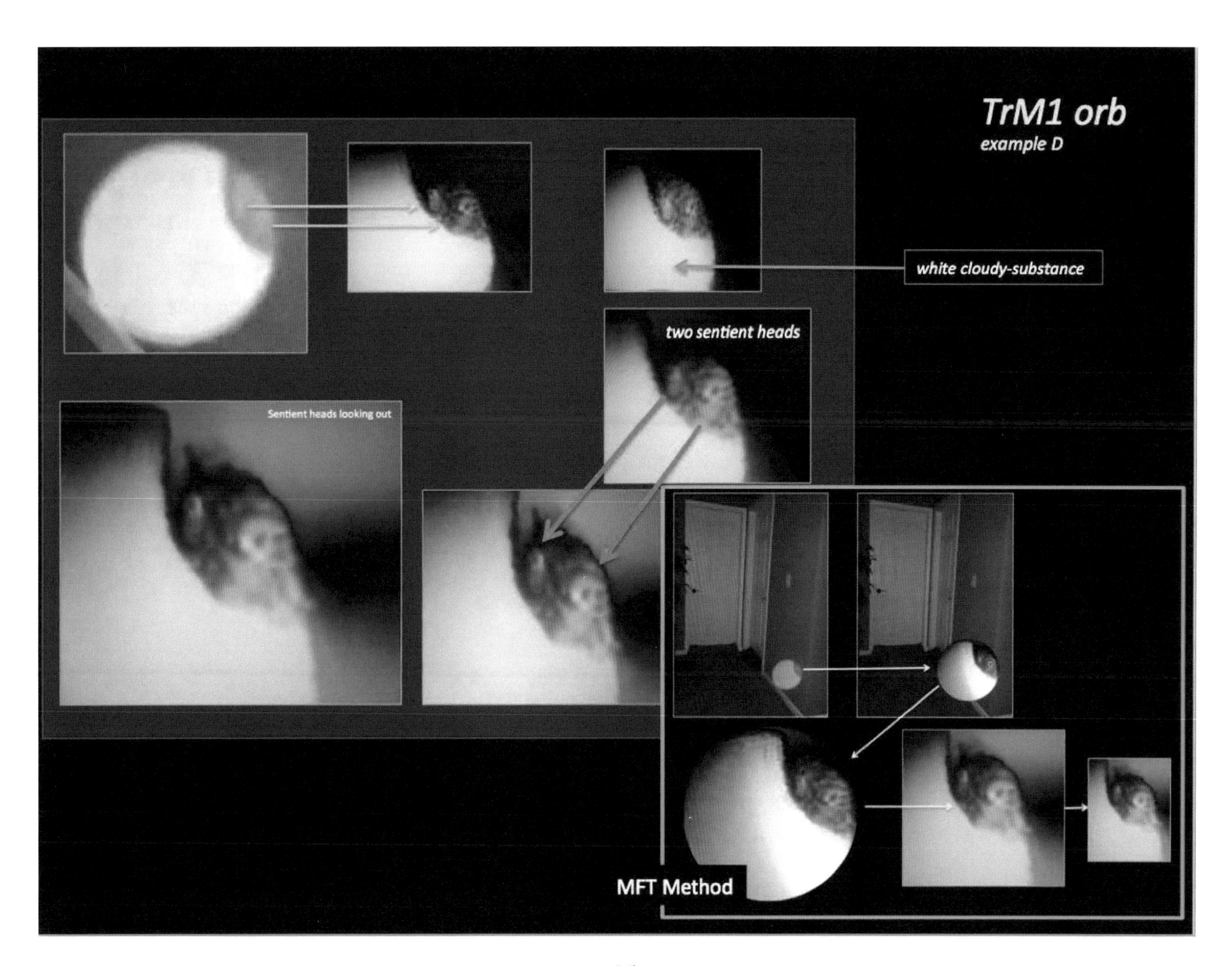
TrM1 orb
example D
white cloudy-substance
two sentient heads
Sentient heads looking out
MFT Method

Part 11
Transcendent Module 2 orbs: (TrM2 orbs)

Module 2 orbs have a lower-key luminosity of brightness, but they can still be easily seen and noticed in photos.
On closer observation when we zoom into them, they seem to have a very thin external membrane. Sometimes the thin external membrane seems to disappear around certain parts of the orb, and sometimes the external membrane may not be noticeable at all.
We will often see Module 2 orbs in various colours of browns, greys, caliginous and dusky shades, and they can also vary in different textures and consistencies. They may sometimes have a smooth outer surface, and sometimes they may have a grainy textured surface.

I believe that the sentient activity and data information inside these modules are from the MID-ASTRAL REALMS of the SPIRITUAL REALITIES. This category of orbs may be for the transportation of the souls, the psyche, spirit embodiment, and the quintessence of incarnation. These are the good souls. There is also another variation of Module 2 orbs, where depending on their missions to heal, to protect or to comfort us, they will transform into a lighter essence, and appear close and around the areas that need to be helped or healed. These I believe are the '***healing orbs***' and will often be seen in a translucent or semi-translucent form of a lighter colour. These variations of orbs will often appear at healing sessions, meditations, spiritual contemplations, and will help to act as a remedial for soothing the soul.
TrM2 orbs may also be the category of orb crafts that come to collect and transfer our soul, when it is our time to pass over to the other side. We will see numerous smaller heads and faces in clusters that often overlap each other, and quite often these beings can be seen with hats or headgear of perhaps the epoch or era of when they where on Earth before they passed over.

GENERAL CHARACTERISTICS FOR TrM2 Orbs:

- TrM2 orbs have a lower-key luminosity of brightness, but can still be easily seen in photos.
- TrM2 orbs often have colours of browns, greys and caliginous dusky shades.
- TrM2 orbs can transform into 'healing orbs' and will change into translucent or semi-translucent forms of a lighter colour, depending on their mission. They may have a thin external membrane.
- The thin external membrane may sometimes not be very noticeable in some areas around the orb.
- Inside the orb, smaller multi-coloured heads and faces can be seen in clusters, and are increasing in numbers.
- Heads and faces can often be seen wearing epoch or era headgear.
- Sometimes we may start to see the occasional upper part of a distorted torso or shoulder.
- Sometimes in TrM2 orbs, we can literally see 'hundreds' of smaller multi-coloured heads and faces.

(See photos, PAGES 97-105)
(Compare TrM1 orbs & TrM2 orbs, PAGES 106-107)

Transcendent Module 2 orb

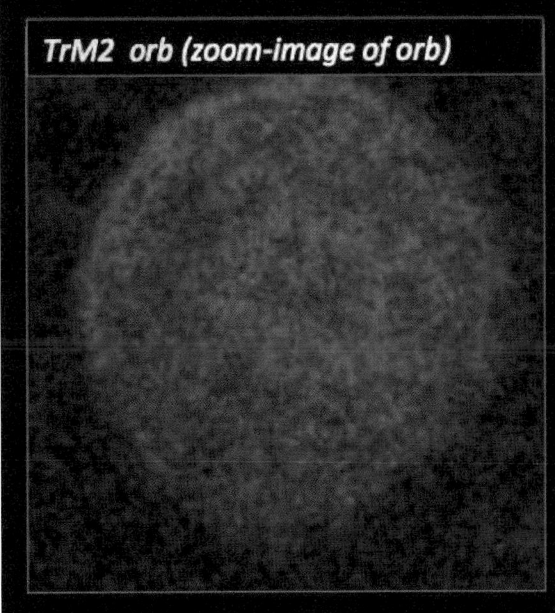

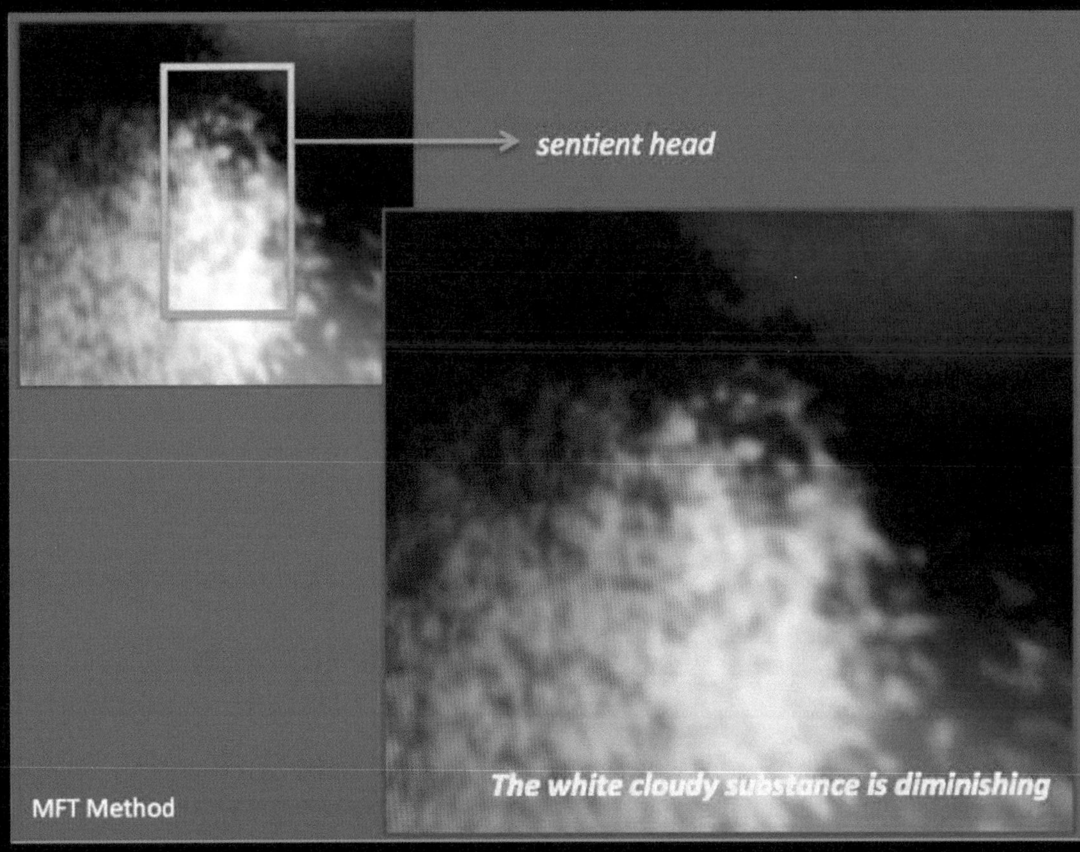

- *No thick aura-membrane around orb*
- *Smaller heads and faces can be seen, often wearing epoch headgear*
- *The white cloudy-substance is diminishing*

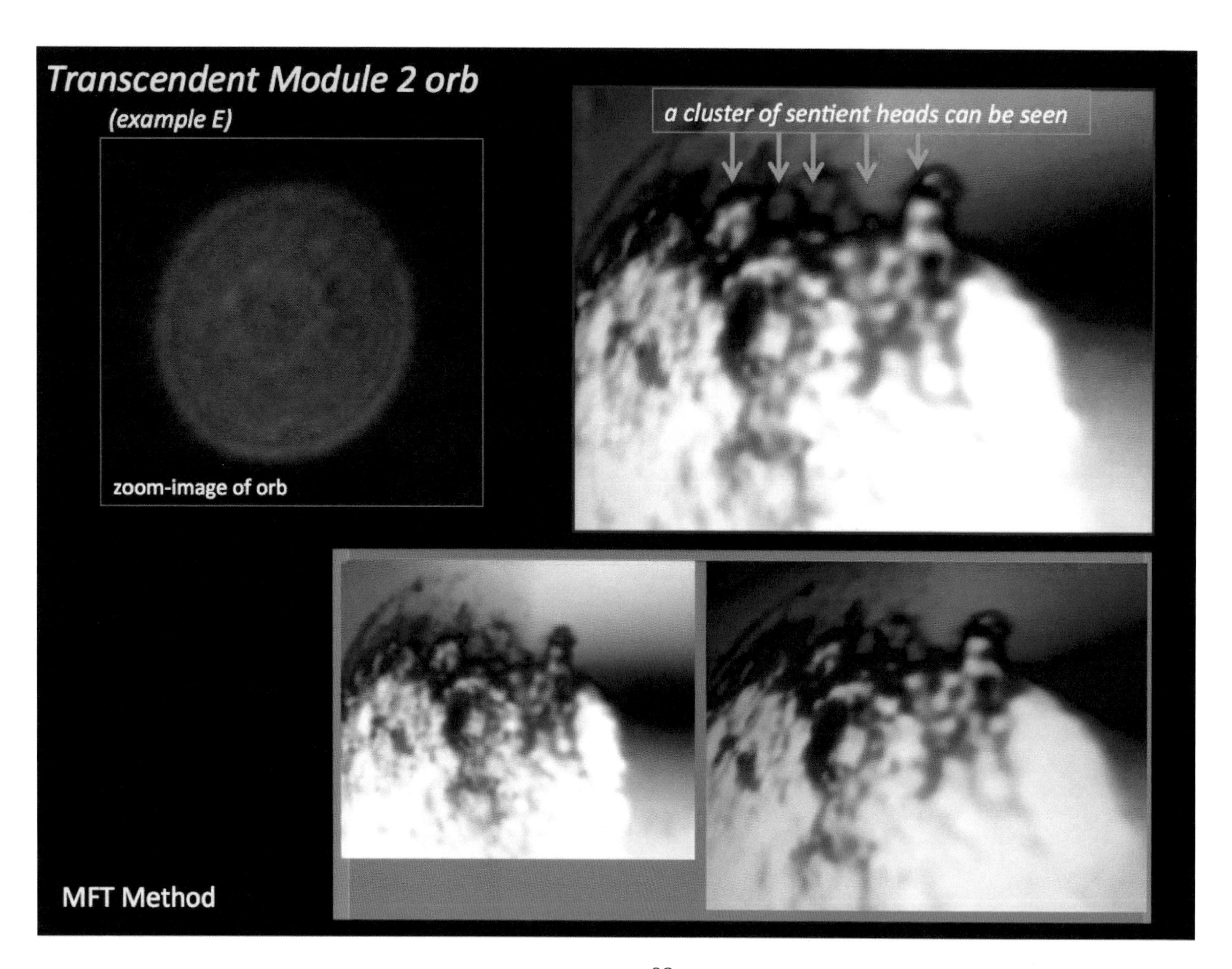
Transcendent Module 2 orb
(example E)
zoom-image of orb
a cluster of sentient heads can be seen
MFT Method

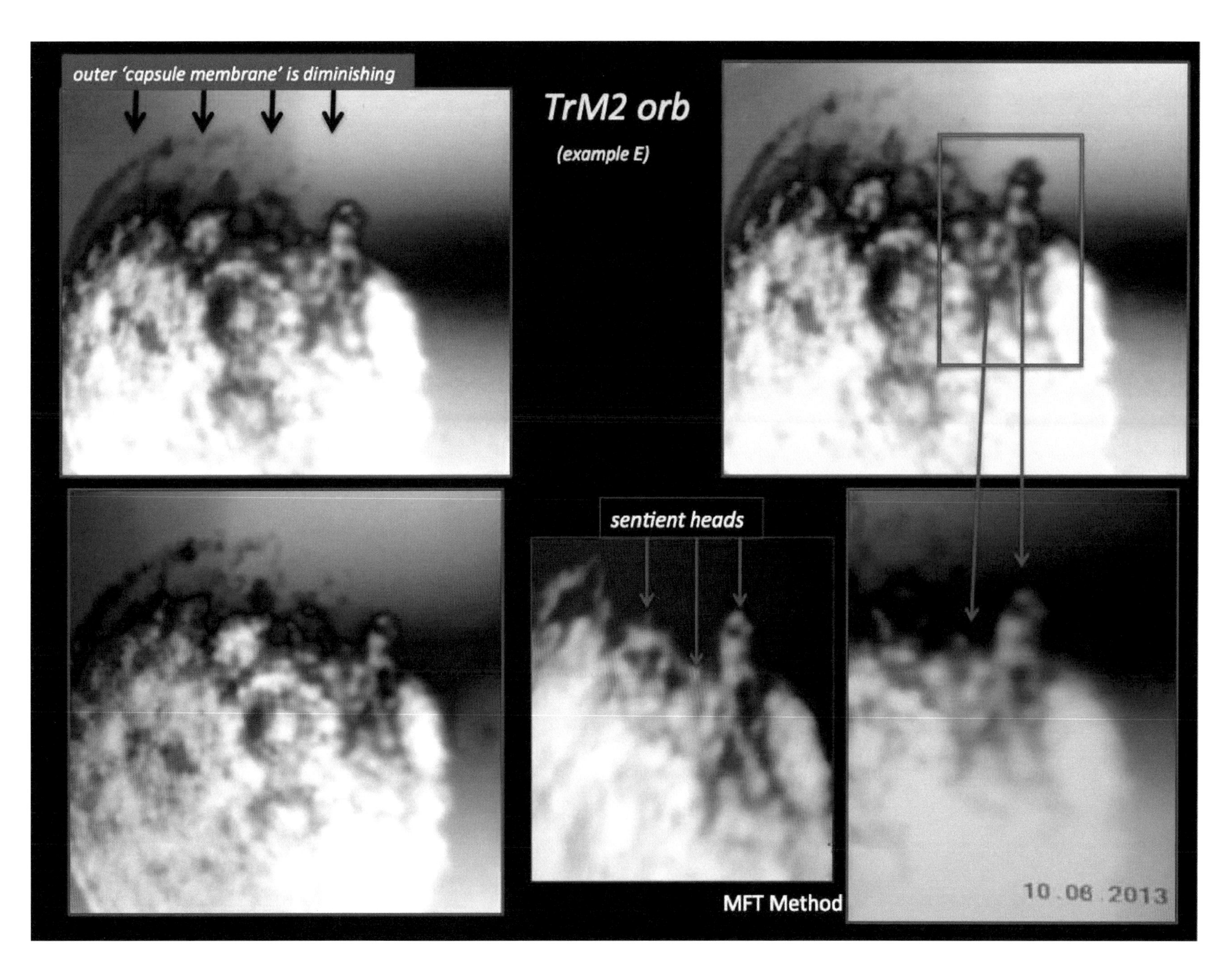

outer 'capsule membrane' is diminishing
TrM2 orb
(example E)
sentient heads
MFT Method
10.06.2013

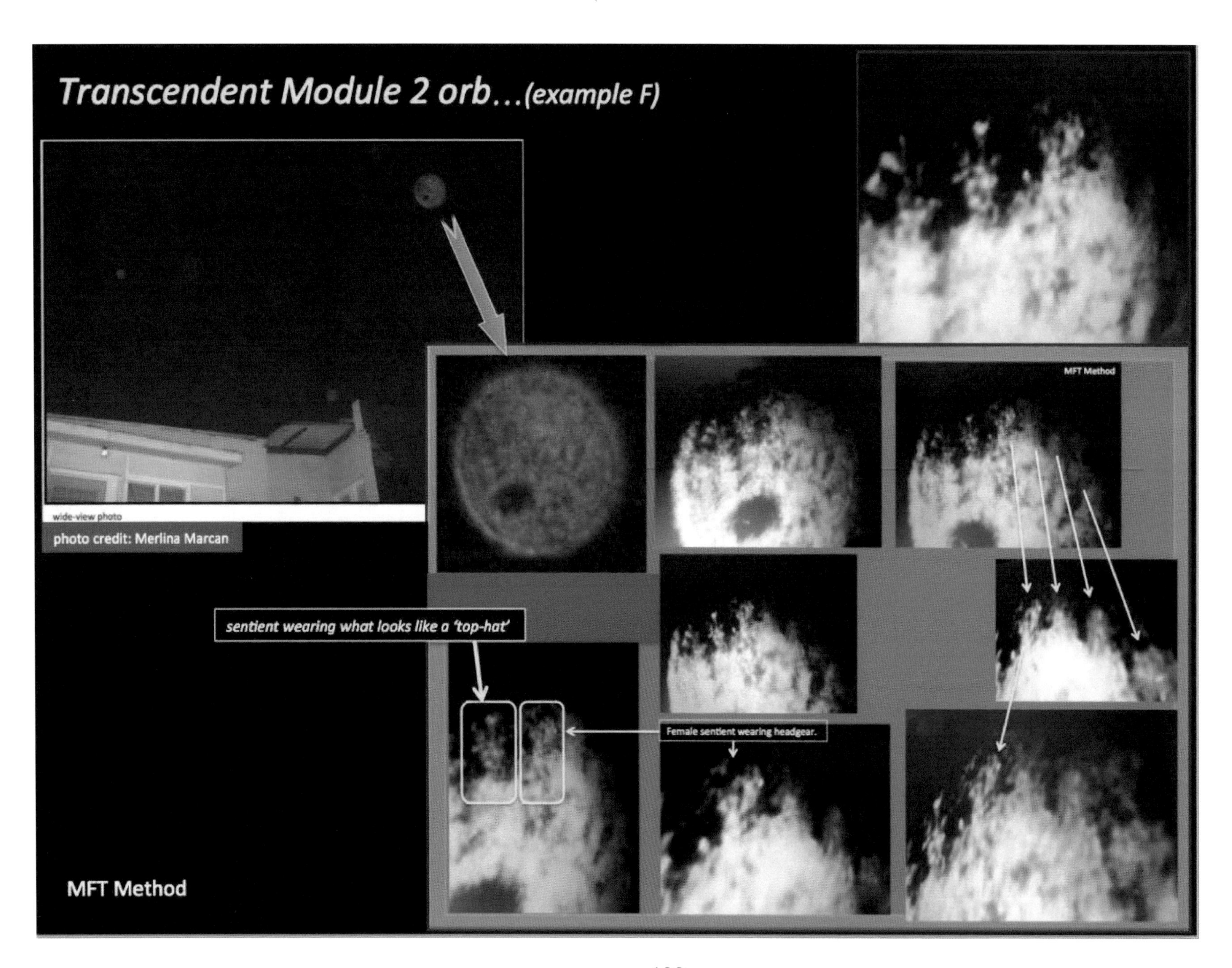
Transcendent Module 2 orb…(example F)
wide-view photo
photo credit: Merlina Marcan
MFT Method
sentient wearing what looks like a 'top-hat'
Female sentient wearing headgear.
MFT Method

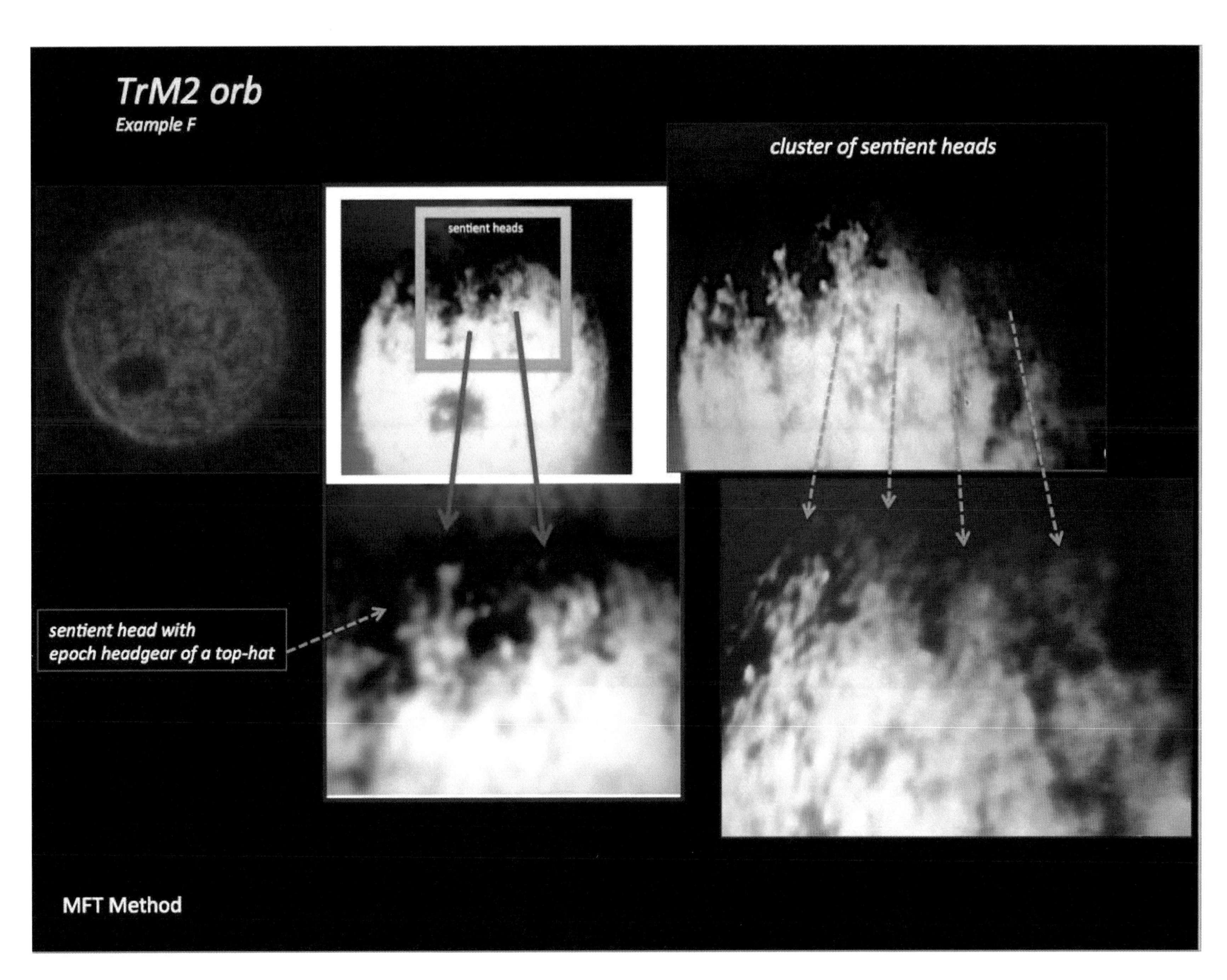
TrM2 orb
Example F
cluster of sentient heads
sentient heads
sentient head with
epoch headgear of a top-hat
MFT Method

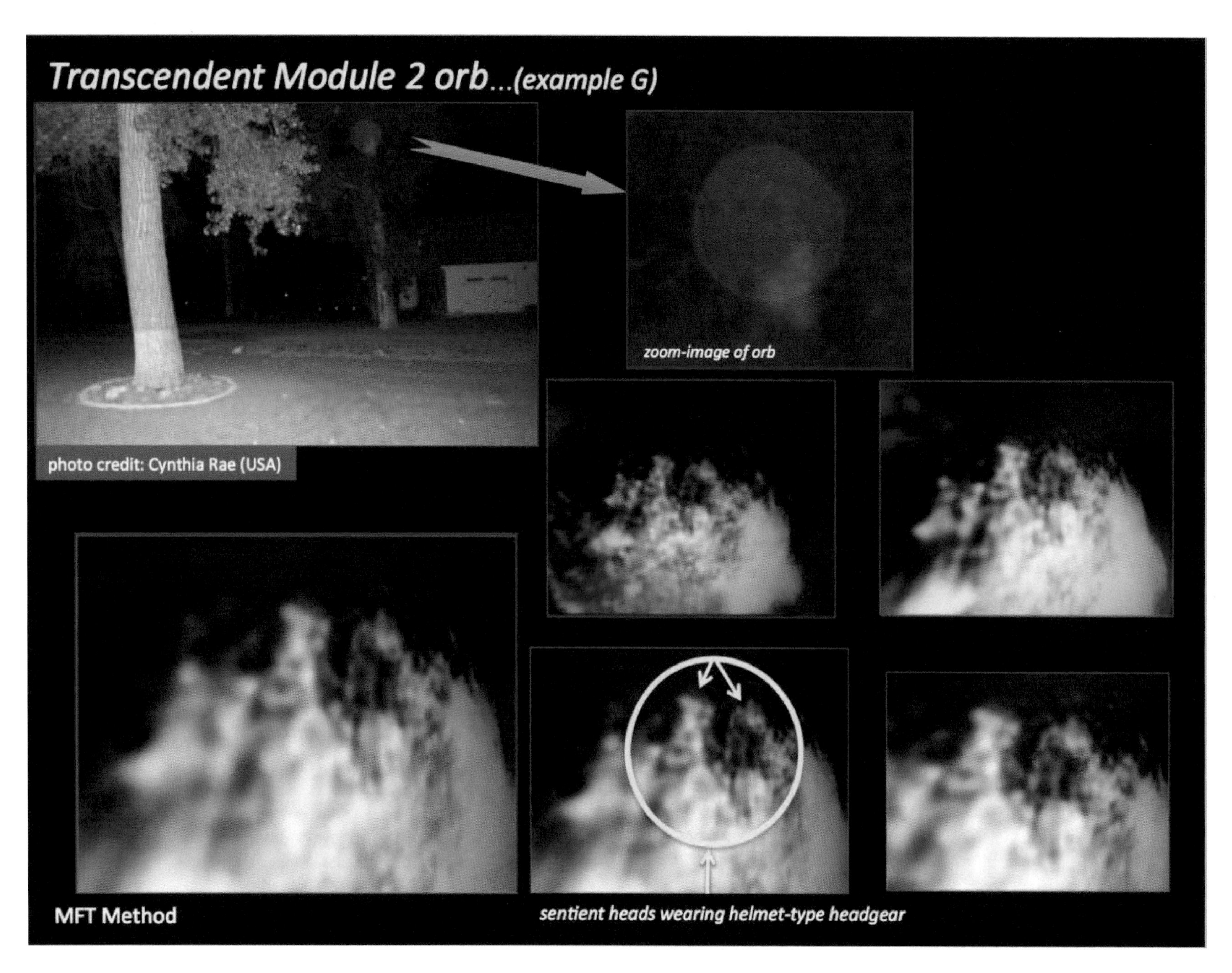
Transcendent Module 2 orb...(example G)
zoom-image of orb
photo credit: Cynthia Rae (USA)
MFT Method
sentient heads wearing helmet-type headgear

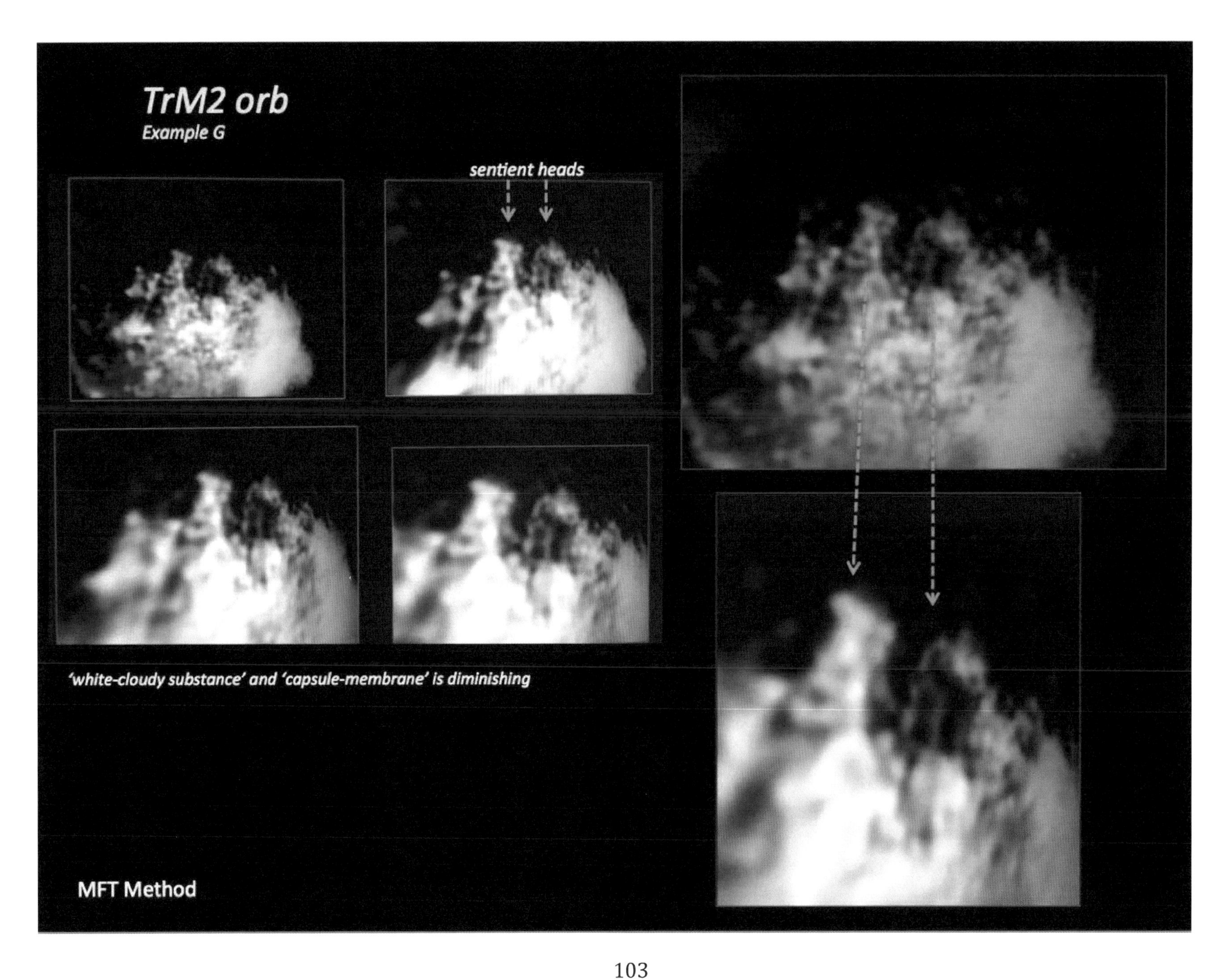
TrM2 orb
Example G
sentient heads
'white-cloudy substance' and 'capsule-membrane' is diminishing
MFT Method

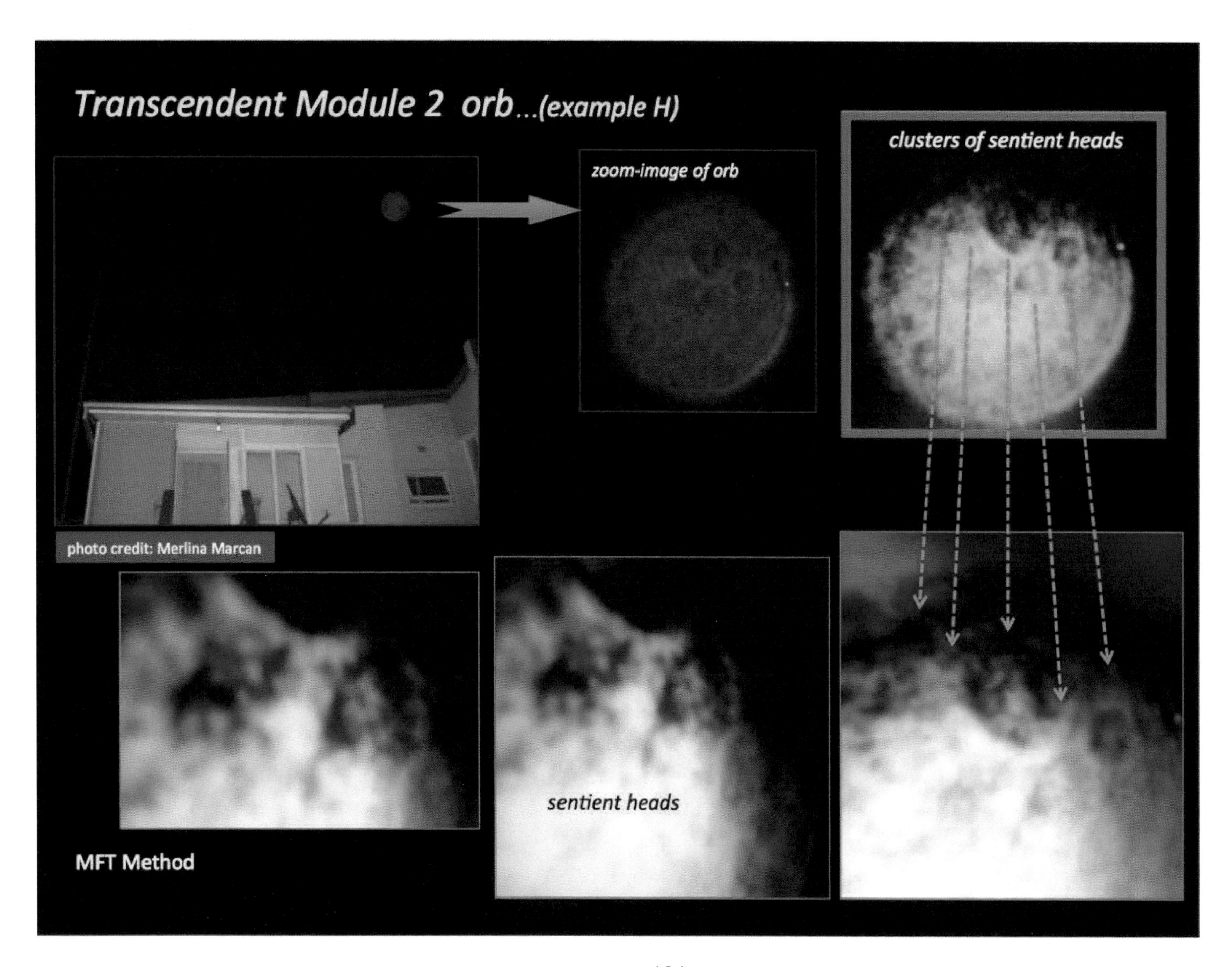
Transcendent Module 2 orb...(example H)
zoom-image of orb
clusters of sentient heads
photo credit: Merlina Marcan
sentient heads
MFT Method

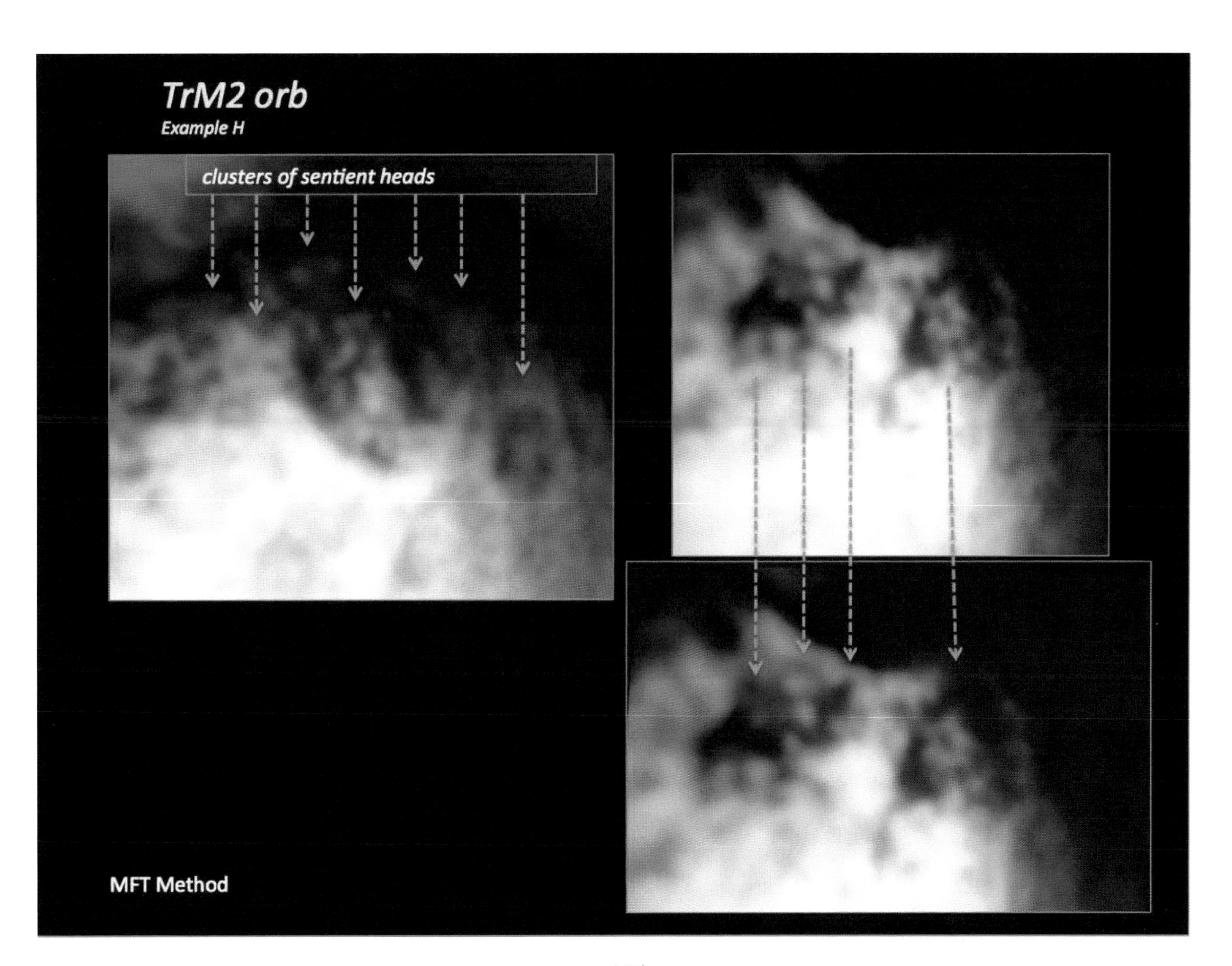
TrM2 orb
Example H
clusters of sentient heads
MFT Method

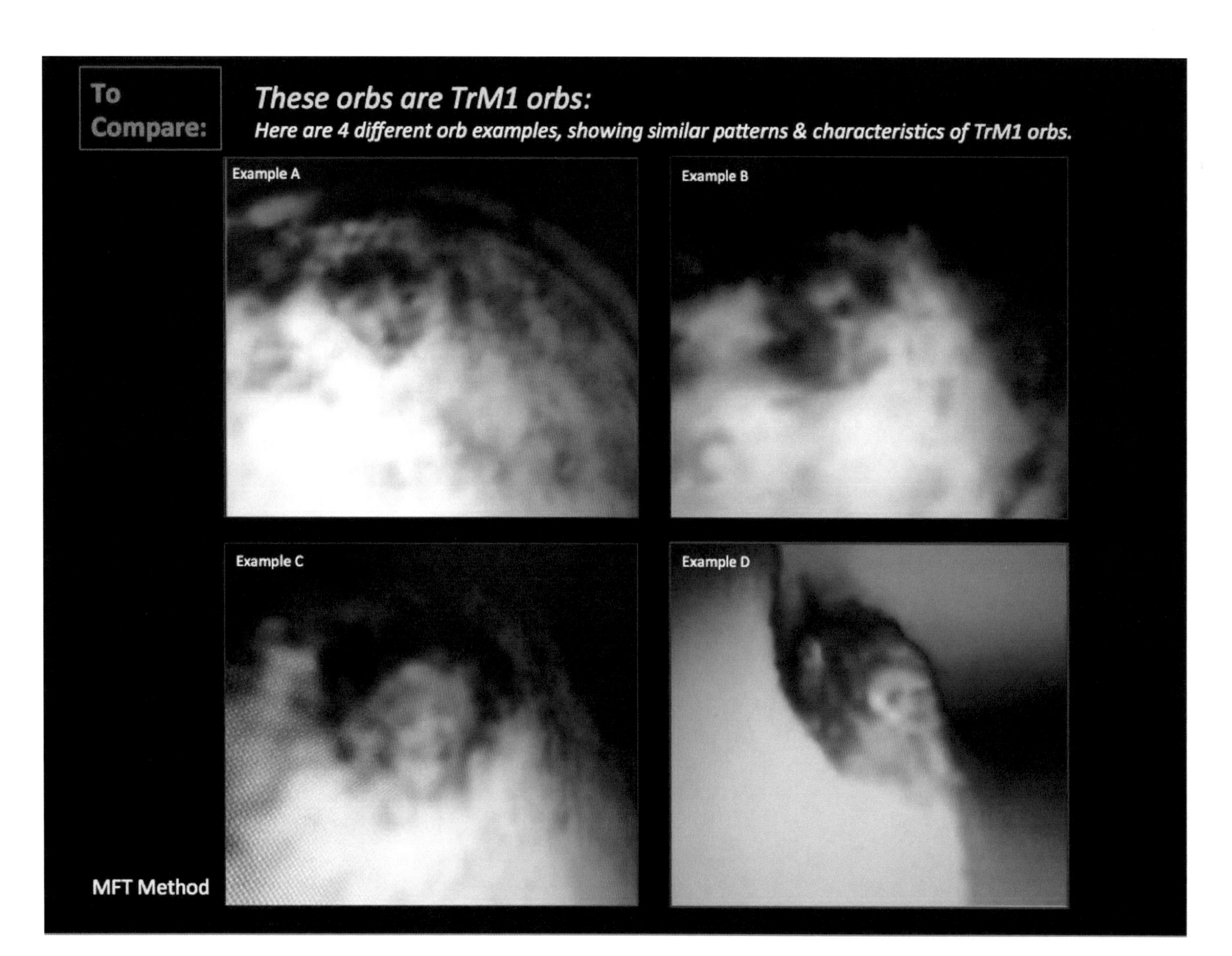
To Compare:
These orbs are TrM1 orbs:
Here are 4 different orb examples, showing similar patterns & characteristics of TrM1 orbs.
Example A
Example B
Example C
Example D
MFT Method

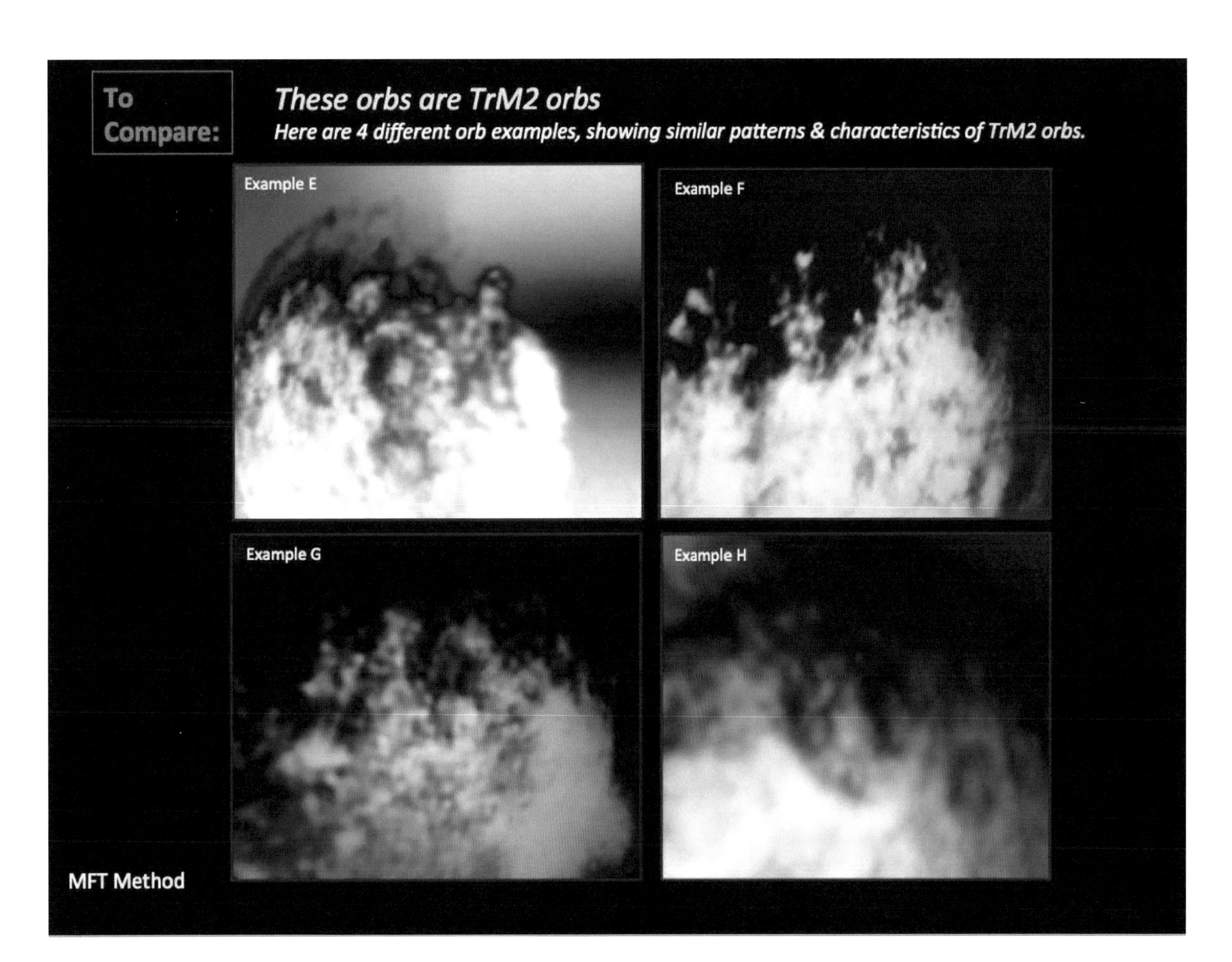
To Compare:
These orbs are TrM2 orbs
Here are 4 different orb examples, showing similar patterns & characteristics of TrM2 orbs.
Example E
Example F
Example G
Example H
MFT Method

Part 12
Transcendent Module 3 orbs: (TrM3 orbs)

Module 3 orbs are the darkest of orbs, and often appear as ***dark smudges in the sky***.
They have no protective outer membrane whatsoever, and have no light reflection, which often makes them hard to spot or see in photos. Nevertheless, even though they are hard to spot, there are more dark Module 3 orbs, but we just cannot see them that easily. In order to see them, we would have to *enhance or brighten* the photo images.

Basically, Module 3 orbs are in total darkness, they have no light reflection and are like dark smudges in the sky. I believe these are the orb crafts from the ***LOW-ASTRAL REALMS,*** and perhaps may carry the souls that must pay their karma.

Sentient activity in this category are *extremely small in size*, they continuously seem to *overlap and inter-twine into distorted shapes*, and are extremely overcrowded, often displaying 'multiple hundreds' of sentient densely packed, and literally '*squashed like cans of sardines'*.
(See photos, PAGES 109-111)

As a comparison, I have added these photos of various 'Crowd Gatherings',
so as to compare and give you a gist of what 'Over-Crowded Sentient Activity' inside
Module 3 orbs looks like.

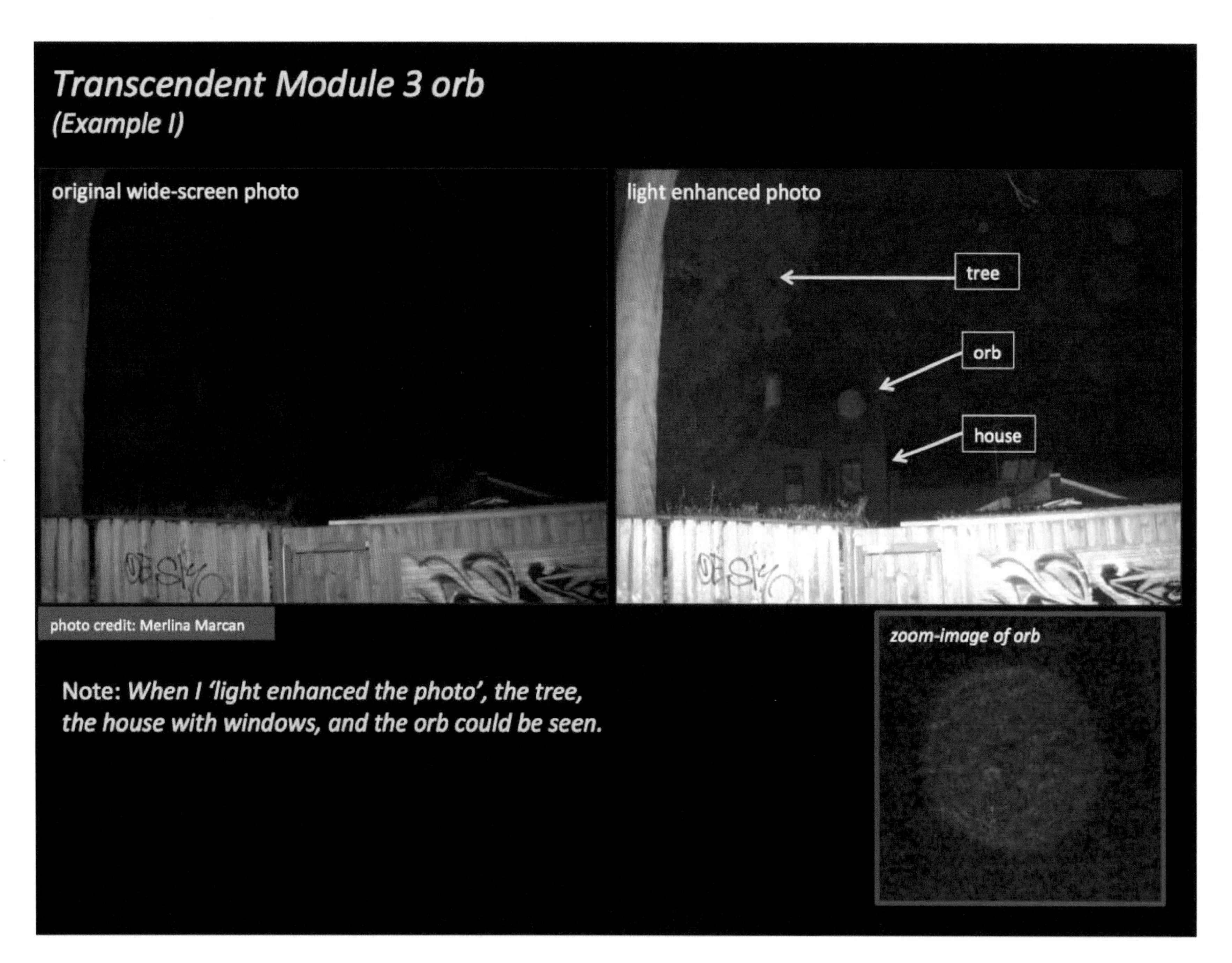
Transcendent Module 3 orb
(Example I)
original wide-screen photo
light enhanced photo
tree
orb
house
photo credit: Merlina Marcan
zoom-image of orb
Note: When I 'light enhanced the photo', the tree, the house with windows, and the orb could be seen.

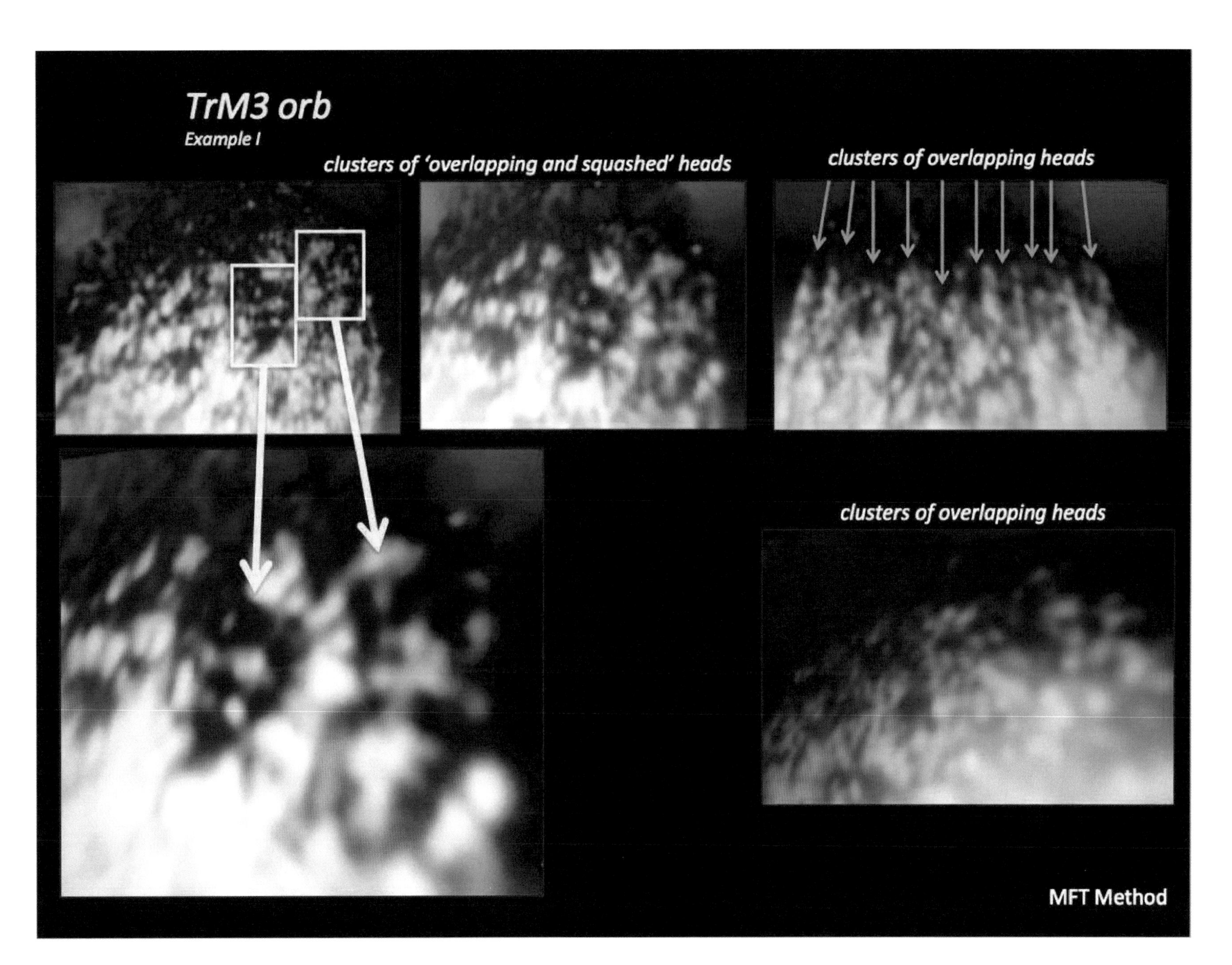
TrM3 orb
Example I
clusters of 'overlapping and squashed' heads
clusters of overlapping heads
clusters of overlapping heads
MFT Method

Part 13
Where are they coming from?

Before I go on to show you some *Case Studies*, I think it is a good idea here, to talk about the different regions or the different areas that orbs may be coming from, because I do not believe that they are all coming from the same region or the same place.
Let us first look at a statement that a former well-regarded Canadian Minister Of Defense has made:

QUOTE:

"UFOs are as real as the airplanes that fly over your head!"

Paul Hellyer (former Minister Of Defense, Canada)

As we humans have *Airplanes and Aircraft* that transport us from one end of the world to the other, or to better phrase it, from one end of the *Earth* to the other, so do 'Other Realities' that can transfer from one end of the cosmos to the other.

Regardless of whether our planes vary larger or smaller, wider or slimmer in shape appearance, our Earthly aircraft all follow same principles of aerodynamics. Aerodynamics that are as advanced as we Earthlings could possibly know about the Fundamentals of 3-D Aerodynamics.

When we see an airplane or aircraft from different Airlines, chances are that we can identify from what Country it belongs to, *by simply looking at little telltale signs such as the planes colours, designs, and the different Exterior Logos*. So…by simply looking at an airplane, we know what country it belongs to.

Note:
The Aerodynamics of Orbs From Other Realities are far more superior to ours, and even though we do not understand how they operate or function, I believe that eventually with more research, we may be able to identify from what parts, what areas or from what regions of the cosmos they are coming from, simply by looking at their colours, designs, exterior hidden logo details and little telltale signs.

There are vast varieties of different 'ORB CRAFTS', coming from 'different regions' of our Galaxy!

(See photos, PAGES 114-120)

What part of 'Earth' do these planes (crafts) belong to ??

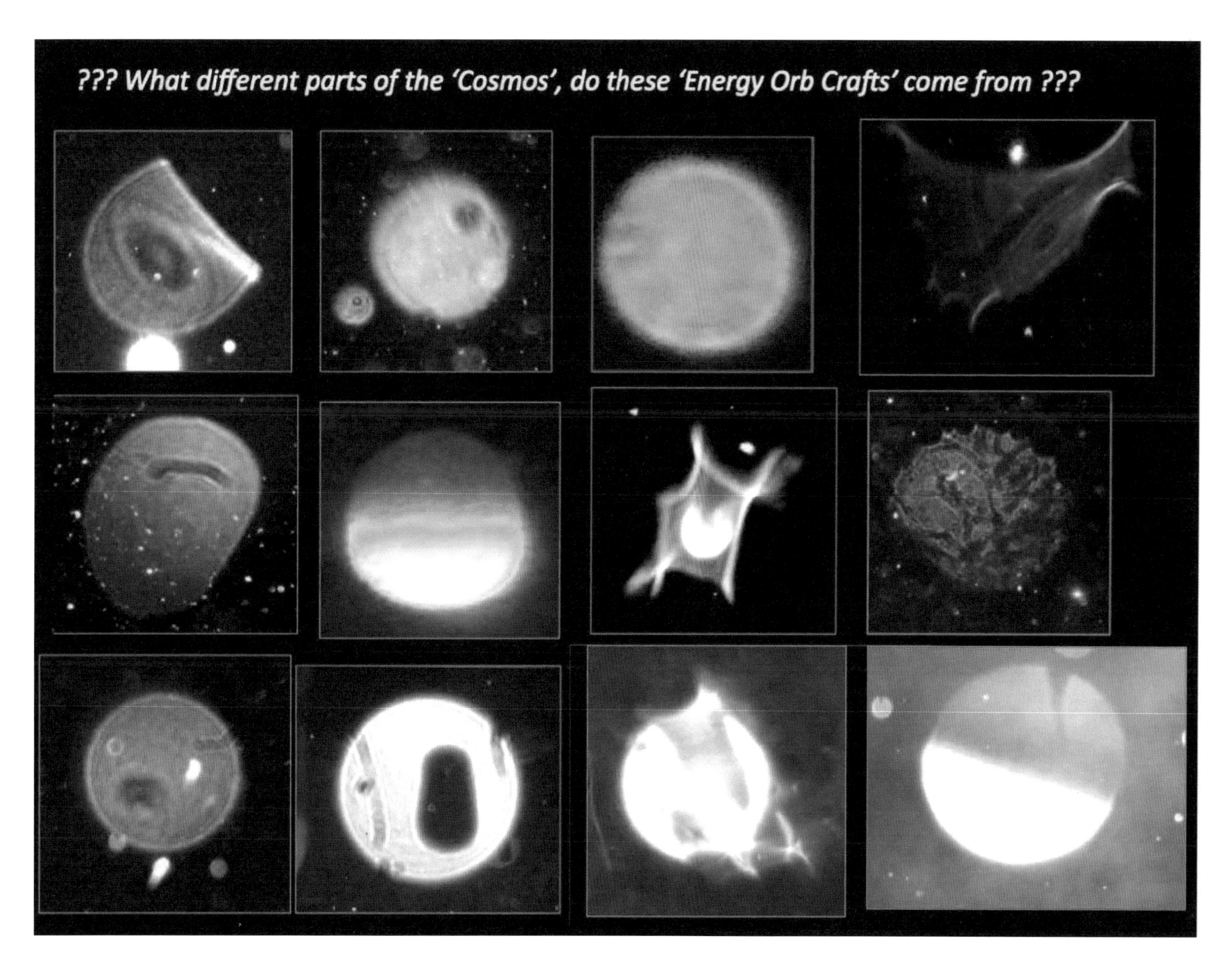
??? What different parts of the 'Cosmos', do these 'Energy Orb Crafts' come from ???

?? Could these similar looking 'Orb Crafts' be coming from same regions or same areas of the cosmos ??

photo credit: Diana Davatgar (Germany)

photo credit: Tracee Gorman (USA)

photo credits: Andrea Corsick (USA)

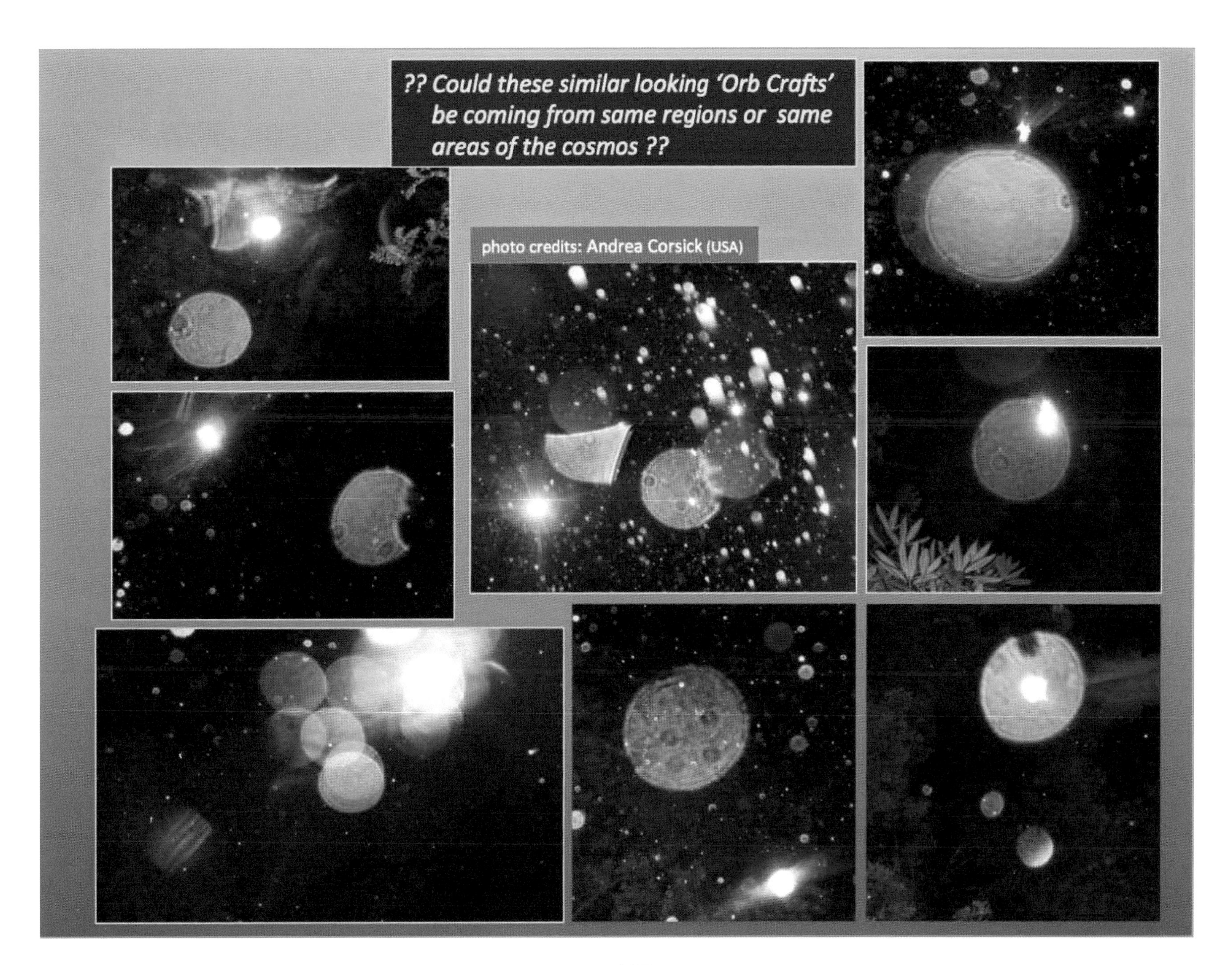
?? Could these similar looking 'Orb Crafts' be coming from same regions or same areas of the cosmos ??
photo credits: Andrea Corsick (USA)

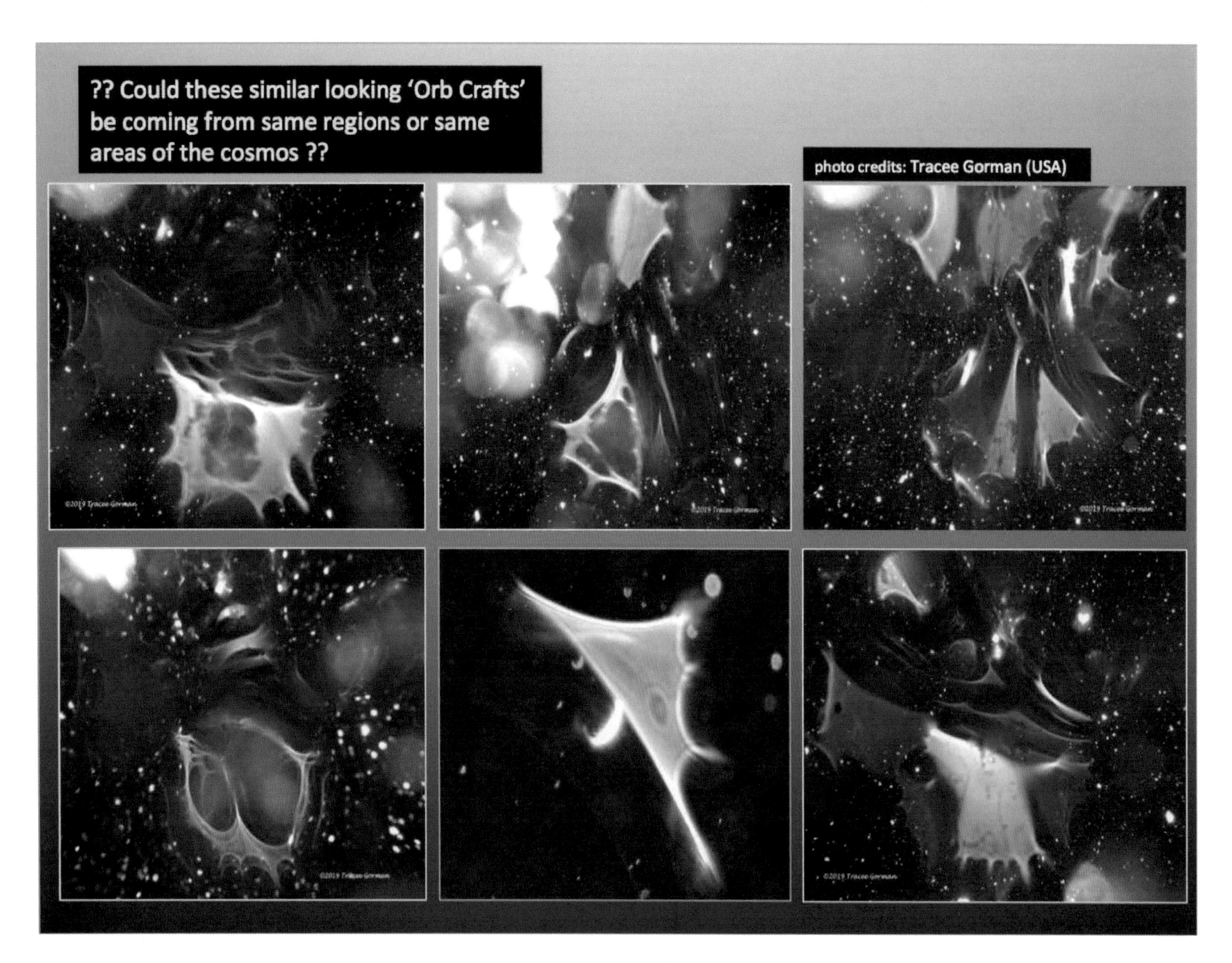
?? Could these similar looking 'Orb Crafts' be coming from same regions or same areas of the cosmos ??
photo credits: Tracee Gorman (USA)
©2019 Tracee Gorman
©2019 Tracee Gorman
©2019 Tracee Gorman
©2019 Tracee Gorman
©2019 Tracee Gorman

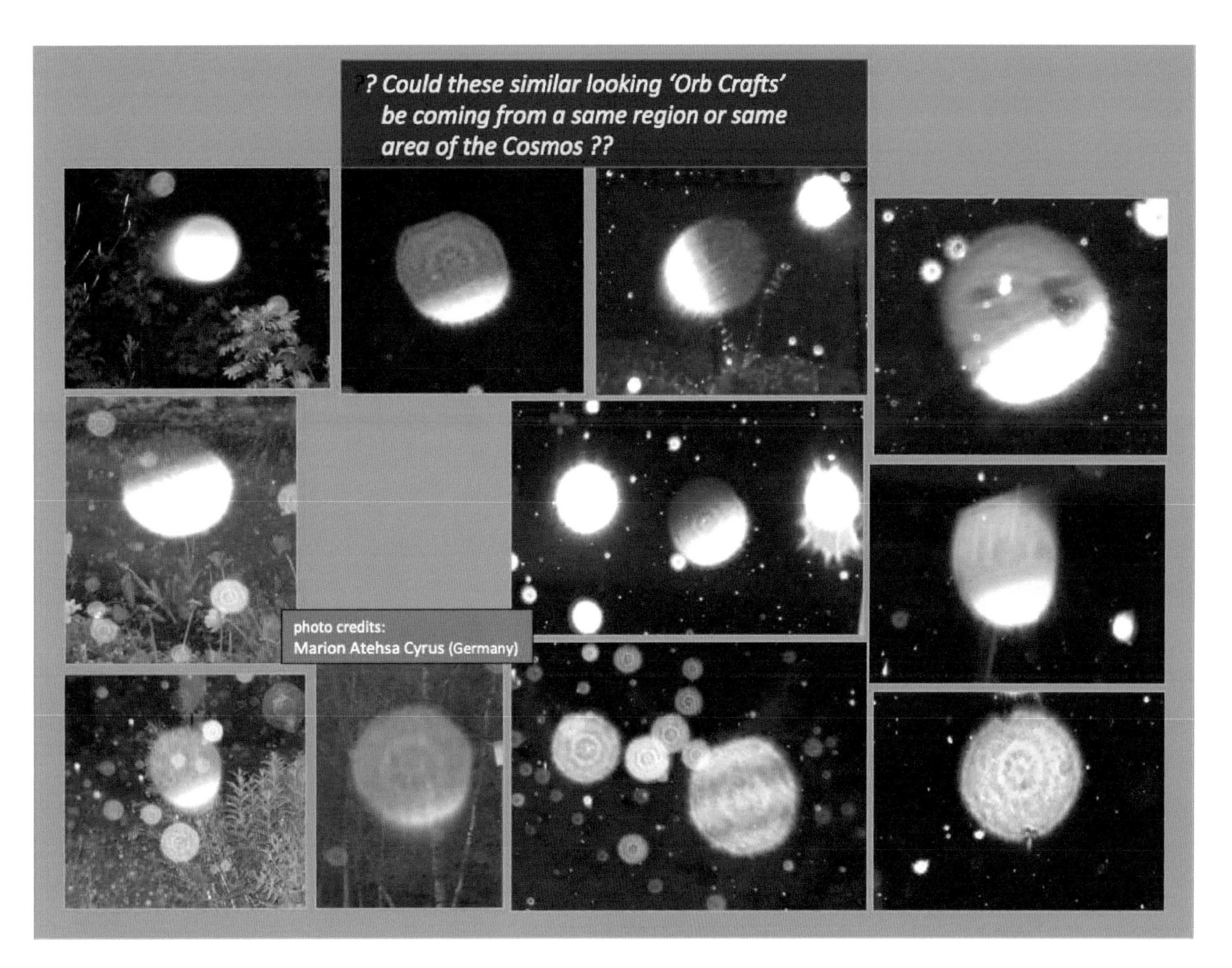
?? Could these similar looking 'Orb Crafts' be coming from a same region or same area of the Cosmos ??
photo credits:
Marion Atehsa Cyrus (Germany)

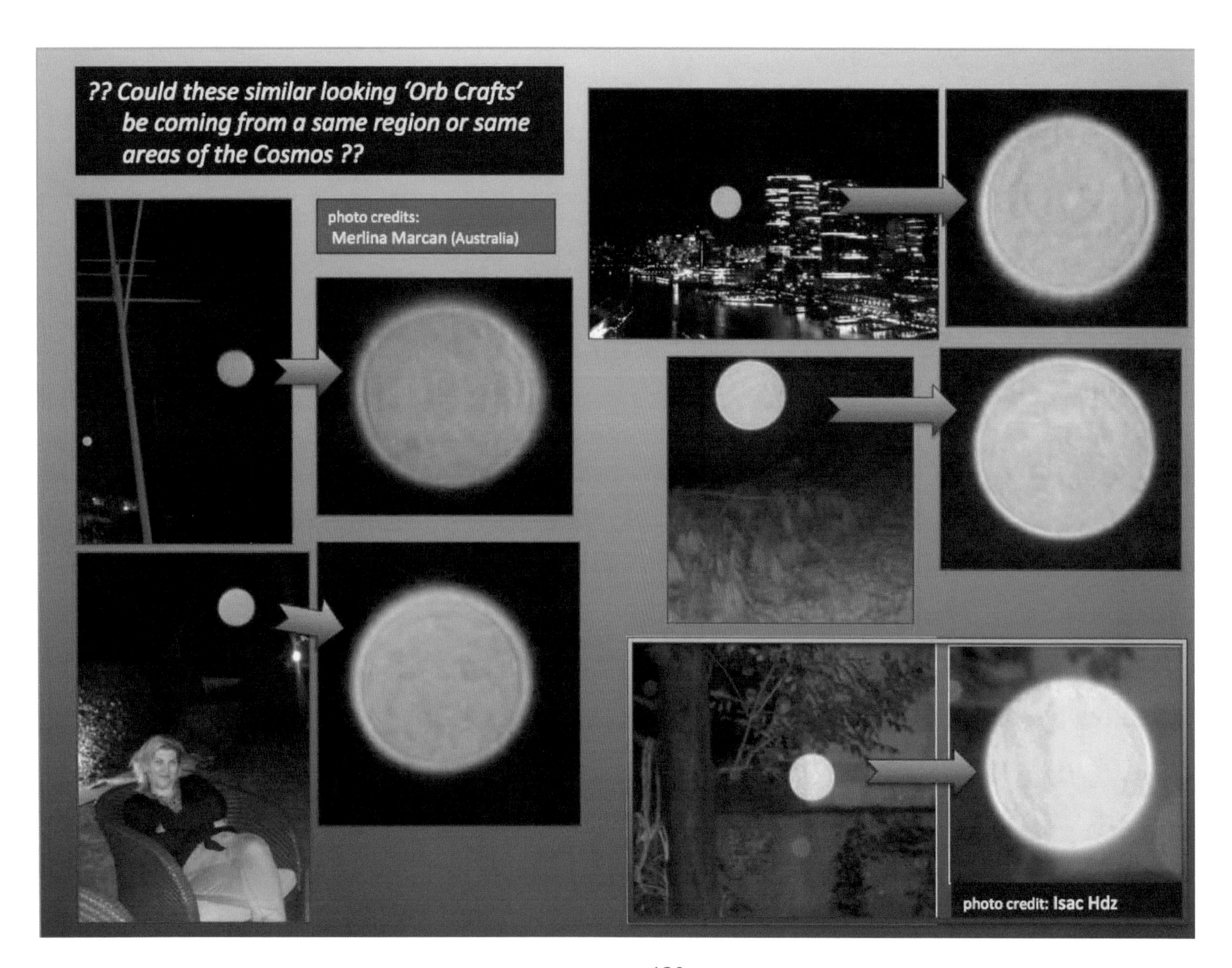
?? Could these similar looking 'Orb Crafts' be coming from a same region or same areas of the Cosmos ??
photo credits:
Merlina Marcan (Australia)
photo credit: Isac Hdz

Part 14
Case Studies

'*Celestial Beings*' have always fascinated humanity (since the beginning of ancient civilizations), and there has always been an ongoing search for signs of Angels, Spirits and ETs. (The term 'Celestial Being' often refers to angels, aliens and other light beings that come from Higher Realms).
We are constantly looking to the skies for signs of other life in the Universe.

An '*Observatory*' is a location used for observing extraterrestrial and celestial events, and one of its main purposes is to explore the unchartered universe and discover the mysteries of the cosmos. Many observatories have been built all around the world, in countries such as Chile, USA, Spain, Australia, South Africa, Puerto Rico, Hawaii, India, and even in the South Pole of Antarctica, to mention a few. In these observatories, scientists use **'Multi-Million Dollar Advanced Equipment, Technology and Telescopes' to try and gather 'visual images', or 'various radio signals' to detect if there are any communication attempts or contact signals, from life beyond our planet.**

They are constantly looking for Interplanetary, Interstellar, and Intergalactic information that may bring new findings. New Scientific Findings have been discovered that the space between things is anything but empty, and the 'Quantum Hologram' of the cosmos is full of a pulsating living essence.

Scientists are trying to focus into the Universe, hundreds, thousands and millions of light years away; their aim is to capture cosmic bodies that are out of our galaxy range. And that's cool! But this is like asking a baby to *sprint* before it can walk! Scientists are searching for answers that may lay *thousands and millions of light years away*, ***BUT...*** they are neglecting answers that may lay a few hundred feet away, or in some cases just a few meters away. The 'Orb Phenomenon' is here, it is next to us and it is close to us.

WE physically cannot travel hundreds of light years to visit THEM, but genuine and authentic manifestations of orbs can pulsate their frequencies and they can come to visit US!

So while scientists are trying to focus into the very far, far away universe, why are they not trying to focus literally into our own back yards so to speak??...Because we need not look any further for confirmation of life beyond our reality… **THEY are HERE**, all around us, and they use ORBS as their travel means.

Over the years of my research, I not only research my own personal orb photos, but I also research orb photos from all around the world. I had often asked myself the question of, ***"Why do some people naturally attract orbs, and others do not?"***

I believe that certain people have certain abilities to attract certain cosmic energies, depending greatly on their own 'inner vibrational energies', their 'aura-emissions' and their 'inner-conscious souls'. This may perhaps contribute to why some attract, and some repel these outside cosmic energies. Some people can naturally *willingly* stimulate orbs to appear in photos, and some people cannot, no matter how much they try!

I believe there are many multiple numbers of '*Other Realities*' that can exist independently of our own reality, and yet somehow these other realities can come within the range of our 3-D space. When these other realities come within our space, they may perhaps be attracted to *some people* for various reasons.
One of these attractions is towards humans that have good, loving and peaceful intentions. Often the light-workers of the world, the emotional-physical and pranic healers, the ascension workers and the peaceful good-doers of the world will quite often attract these 'beautiful beings of light' that are on a mission to bring peace and love to our world.

Humans and all living creatures possess an *Aura of a Bio-field*. The bio-field takes its power from that living life force whether it, be *people, animals, plants and sacred places*. The light beings that come to visit us, may be attracted to some of these Aura bio-fields and therefore they will often stop their motion, become perfectly still (perfectly round in shape), and pose for the photos, so that we may *see them!*

These celestial beings are Beautiful and Powerful Beings. They are guardians of humanity, and they are the protectors of the World and the Cosmos! When we capture these higher dimensions in our photos, we may feel a true angelic experience, we feel warm, safe, protected, loving and comfortable.

(As I said before…dust, moisture and other contaminations can sometimes produce similar round-like spotty false orb structures in photos…**BUT**…*dust cannot think*, nor can it act in an intelligent way, it cannot strategically position itself in photos to send us messages, and *dust certainly will not come forth by mentally asking it to do so)!*

THE ORB PHENOMENA is casting a new light on the Spiritual and Scientific questions of whether other dimensions can co-exist beyond our material world, and whether they can come to visit us!

According to a new study, the observable universe is made up of at least *two trillion galaxies*, and each galaxy may come in different shapes and sizes. A galaxy may also be called a star cluster or 'Star System'.

I believe that we are often being visited here on Earth by '*star travellers*' from near-by star systems that are thousands of years more advanced than us, and their means of transportation through the cosmos are the super high-tech aerodynamics of 'Orb Crafts'.

So whether we call them star travellers, extraterrestrial travellers, cosmic beings, inter-dimensional, spiritual beings or celestial beings, some people here on Earth may have *special connections* or special bonds with specific areas of the cosmos, and they will attract certain types of *energy fields or orbs* as compared to others. It is like there is a soul-connection on a higher level of consciousness with some people, and we can see this with visual images of what type of 'energy fields' or 'orbs' they most often attract.

(I have decided to show you '7 *different Case Studies'* that I have put together, so that you can see for yourselves how different people and different sacred areas, or vortex-portal areas can sometimes attract different types of 'Energy Fields' or different types of 'Orbs' from different *Areas or Regions* of the Cosmos).

CASE STUDY 1:

Diana Davatgar (Germany)

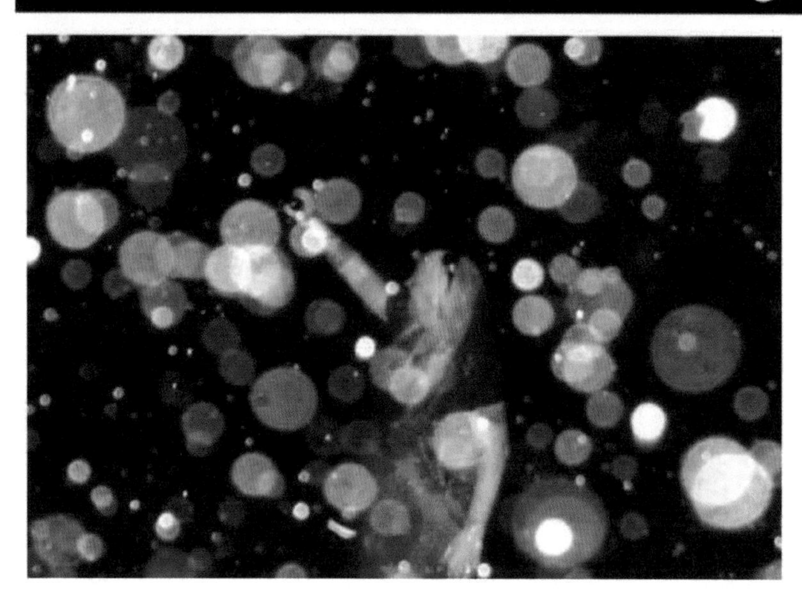

Diana often attracts clusters of round orbs, with many different colours when she is 'outdoors'.

But when she is 'inside her house', she often captures bold, fast flowing 'ribbons of energy fields' in many vibrant colours, such as bright blues, yellows, orange and bright pinks.
I believe this is quite a rare phenomenon because we do not often see these types of 'colourful bold ribbons of energy' very often to come on command.
Perhaps Diana has a special bond, or a special cosmic connection to this rare type of colourful energy fields of consciousness. ***(See photos, PAGES 125-127)***

bold, fast flowing 'ribbons of energy' in many vibrant colours inside her house

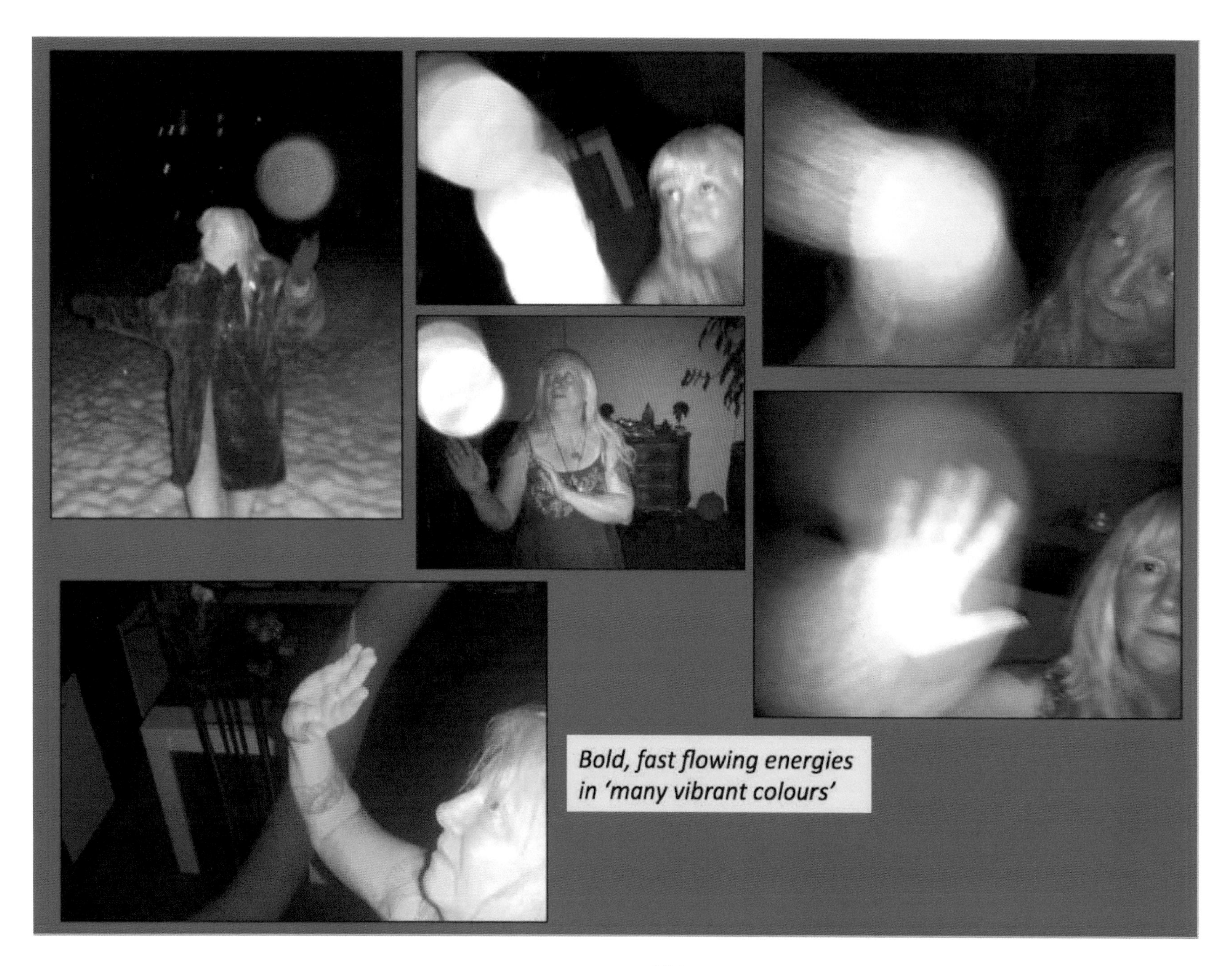
Bold, fast flowing energies
in 'many vibrant colours'

CASE STUDY 2:

Peter F Kahrs (Norway)

*Peter sometimes attracts '**mists of plasma**' that seem to engulf him completely, while he is meditating.*

*But most often, he consciously connects with the '**brilliantly bright whites, silvery and golden large orbs**' that are perfectly round, meaning that they have stopped their motion to a stand-still to surround him and be close to him, as he enters a trance state of enlightenment.*
*I also noticed that Peter attracts orbs that show signs of them '**shape-shifting**', and often we will see orbs near and around him that are '**opening-up**'. Many of the orbs around him can be seen opening-up, similar as a comparison to portal windows on an aircraft. (These openings often look like parts of the orb is missing or may even look like a damaged orb, but in fact these areas are opening up windows which I call 'dark stigmas or patches' in orbs, and these areas very often show sentient heads looking out of these windows).*
Peter is a shaman, a lightworker and healer who attracts energies from the Higher Dimensions that want to bring love and peace to our world. ***(See photos, PAGES 129-132)***

'Plasma mists' surround Peter in meditation

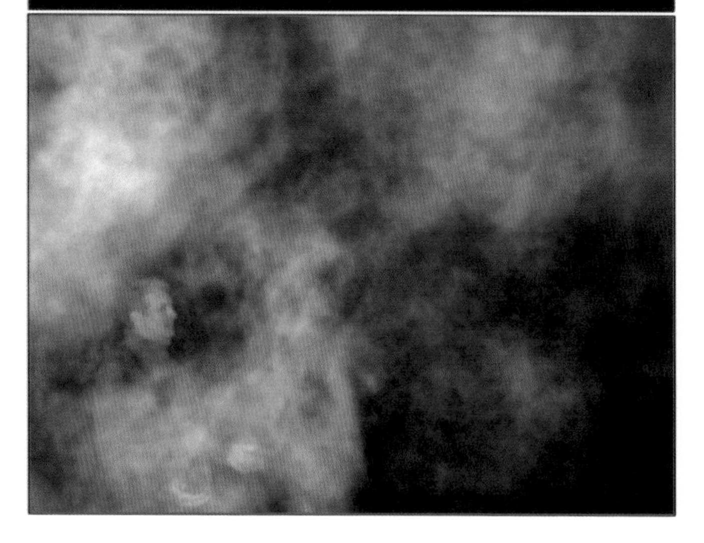

Perfectly round opaque 'zero motion orbs', in brilliant whites, silvers and golden colours.

brilliantly 'bright whites', silvery and golden, large shape-shifting orbs

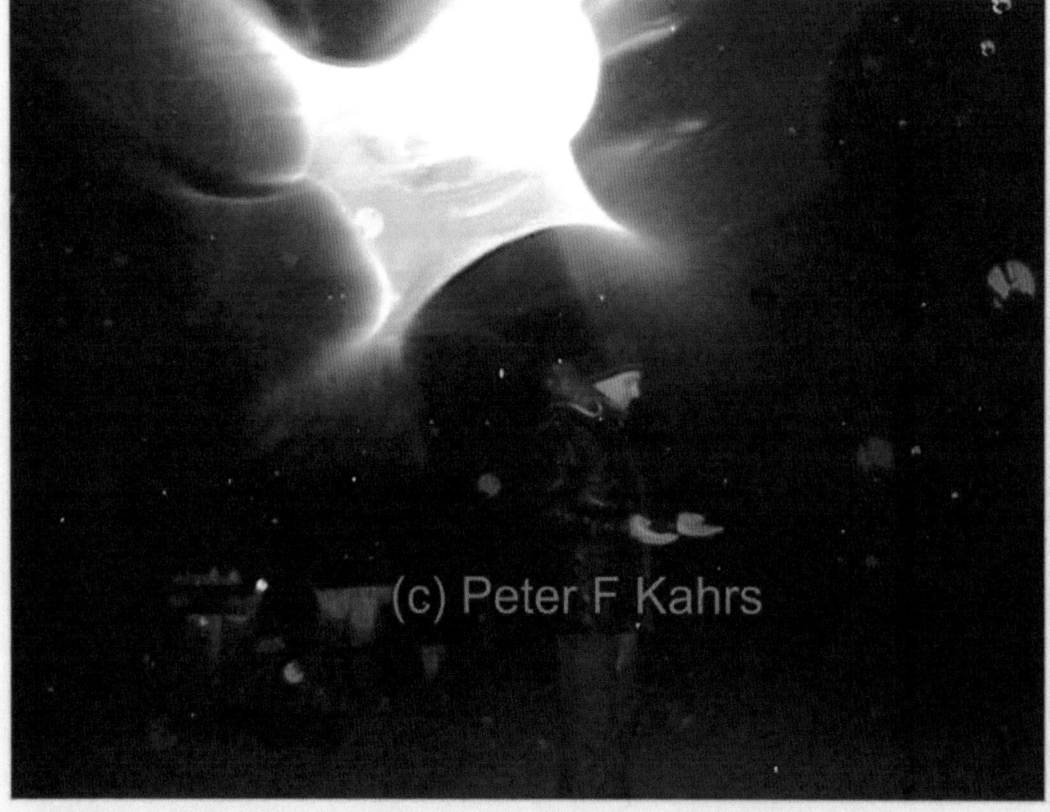

(c) Peter F Kahrs
(c) Peter F Kahrs
'Opening-up orbs' in transformation

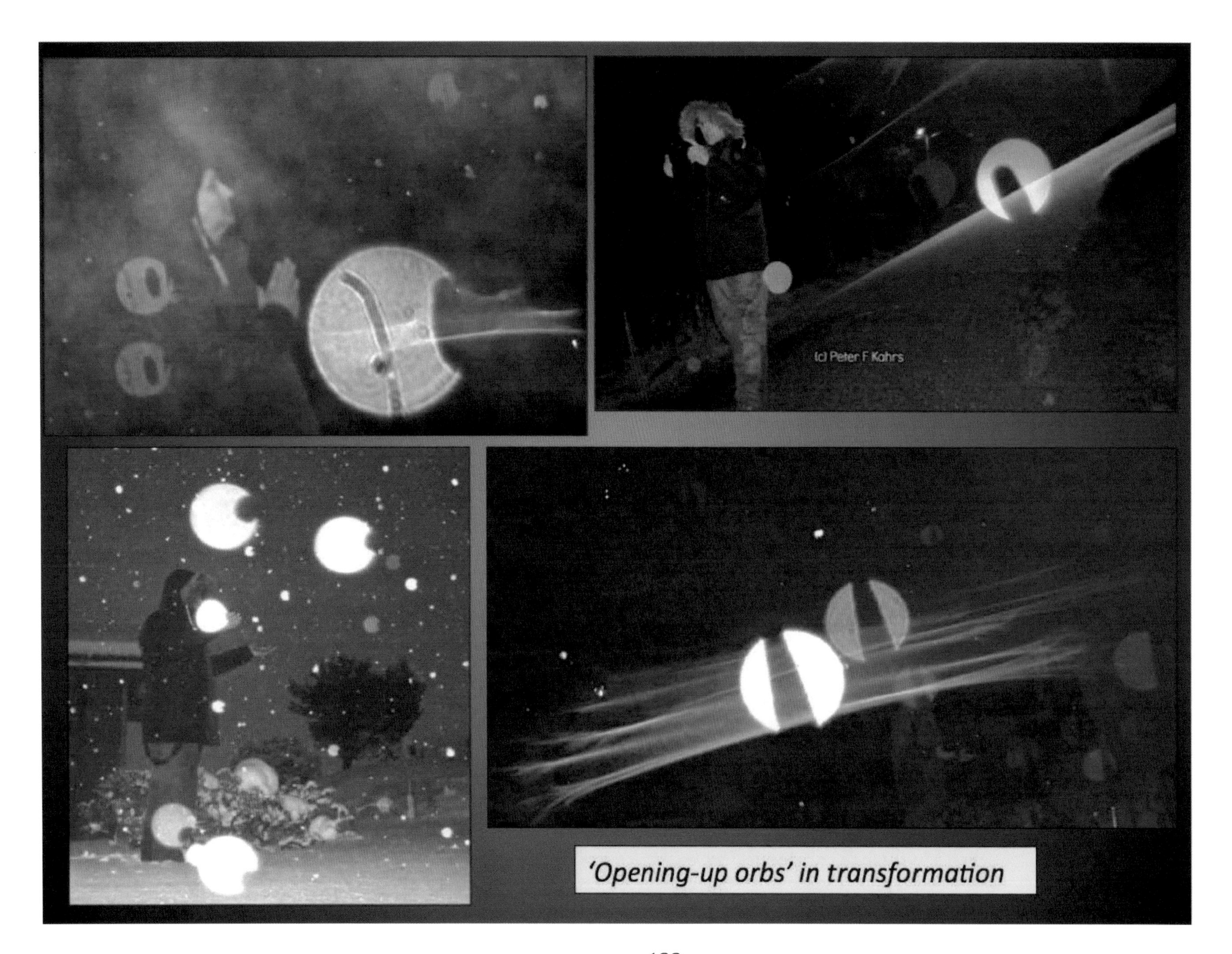

'Opening-up orbs' in transformation

CASE STUDY 3:

Cathy Finch (UK)

Cathy is a lightworker, ascension reiki and pranic healer who also captures bold and brilliant white orbs that surround her when she meditates and prays for peace, health and love on Earth.
*I noticed that Cathy very often attracts **perfectly round orbs**, that have stopped their motion to be near her. Especially when she is engaging in her 'healing and therapy' sessions, the orbs are mostly **'semi-translucent and transparent'**, which I believe these types of orbs may be the **"Healing Orbs"**. They **strategically position themselves** near and around the areas of the bodies of the people that she is concentrating on when she is doing therapy healing.*

Please note:
*Both **'Cathy Finch'** and **'Peter F Kahrs'** are both lightworkers for the good of Earth, they both do meditations and healings, but very often, the types of orbs that they attract are different genres of orbs, which may perhaps explain that they may be attracting **different regions of the cosmos** in helping them do their work.*
(For example, Peter attracts many orbs with 'opening-up windows and many shape-shifting orbs', whereas Cathy mostly attracts the perfectly round semi-translucent ones with no shape-shifting openings). ***(See photos, PAGES 134-137)***

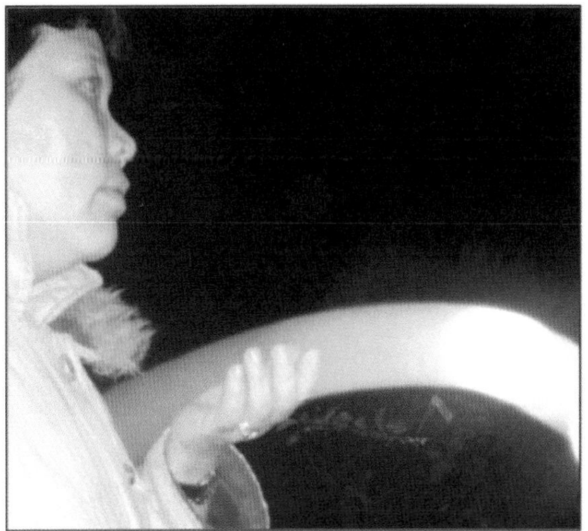

Perfectly round 'zero-motion orbs'

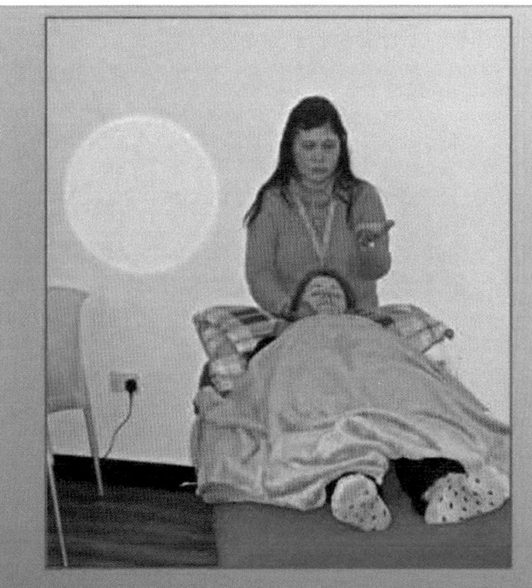

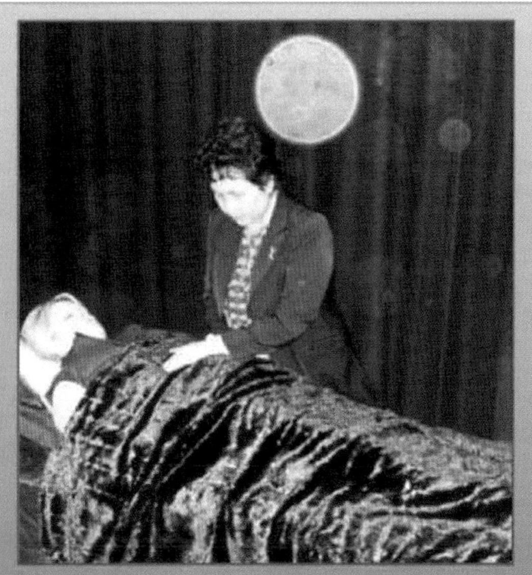

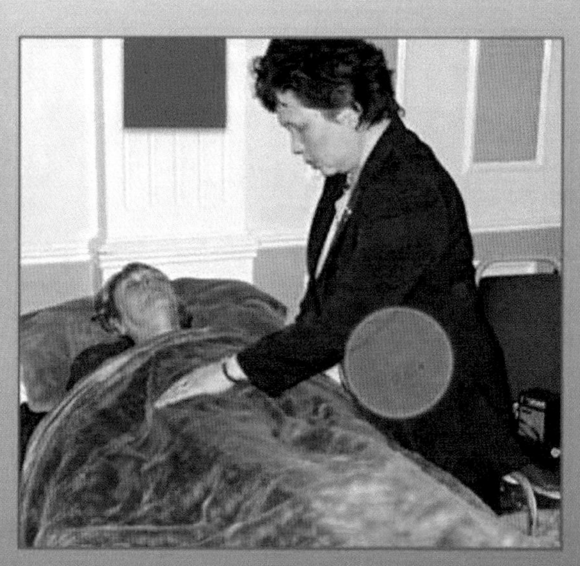

Orbs 'strategically position themselves' in healing sessions

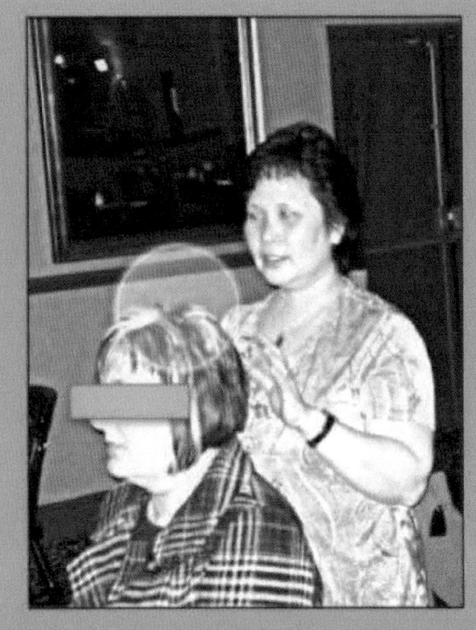

Perfectly round 'semi-translucent orbs' that are strategically positioned for 'Healing'

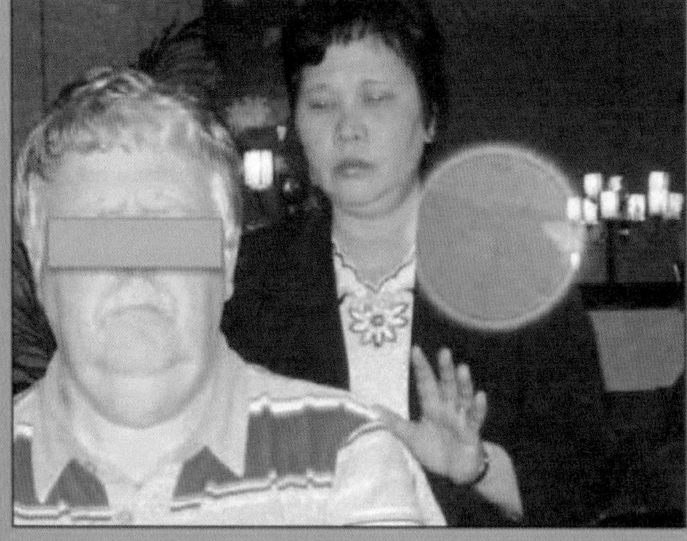

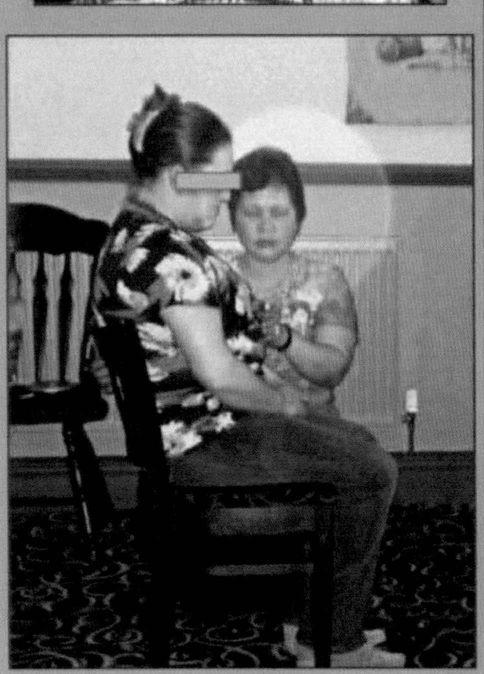

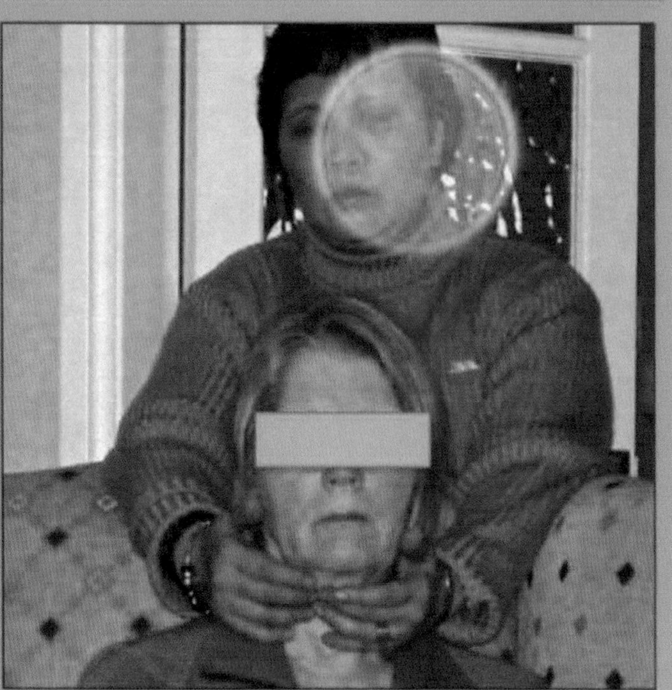

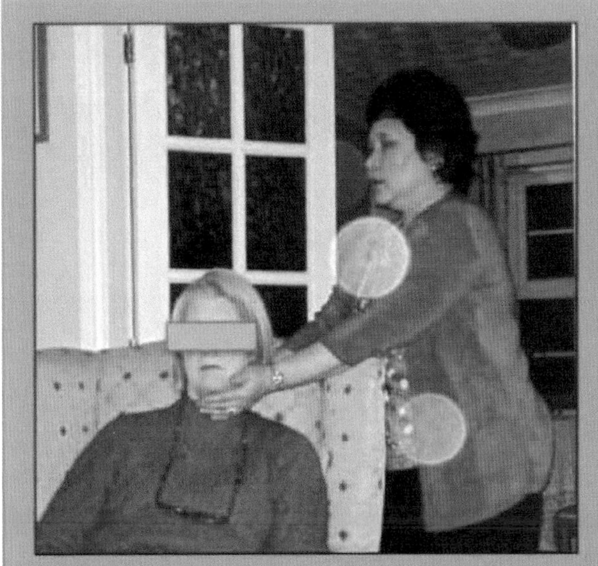

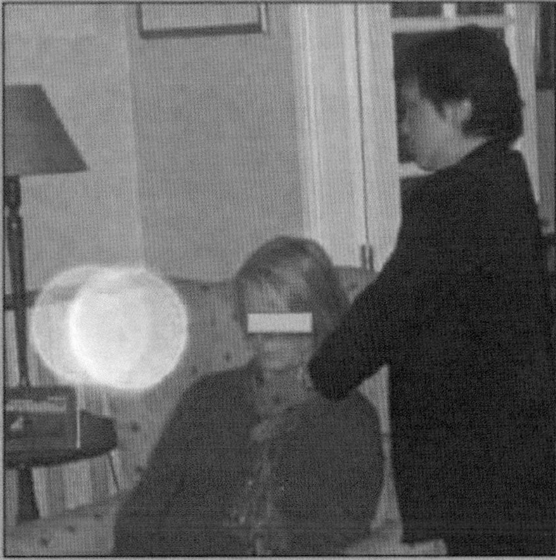

Perfectly round, semi-translucent orbs that are strategically positioned for 'healing'.

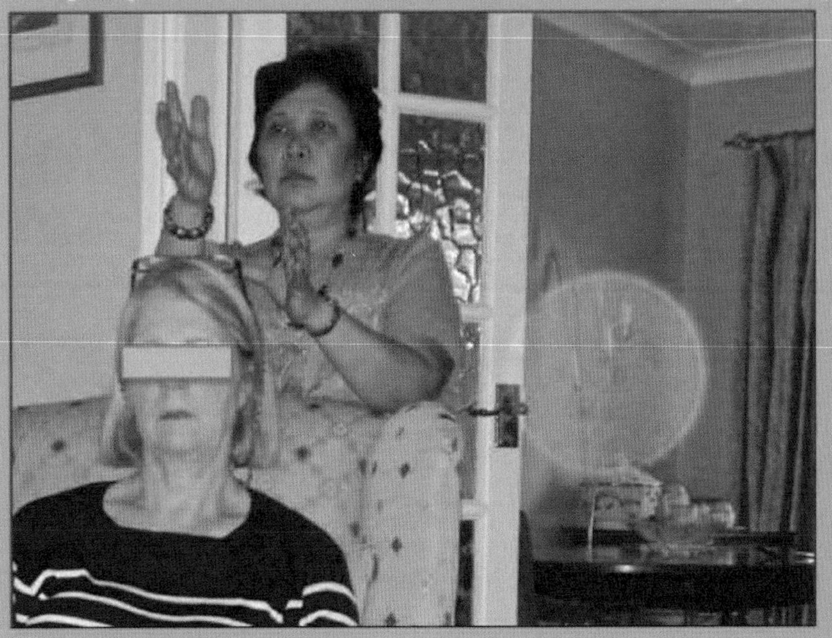

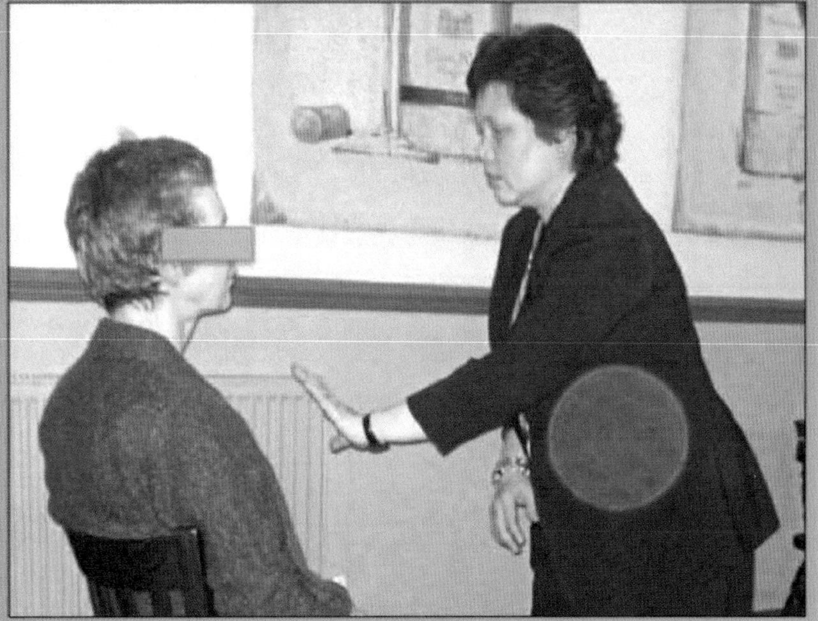

CASE STUDY 4:

Foco Tonal (Hunab Ku)
Aguascalientes City, Mexico

Special thanks and credits to:
Graca Pimentel (USA)
Juan Ishtar Abdiel (Mexico)
Maratt Sandoval (Mexico)

Portals and vortices exist more abundantly than we are aware of.
*A '**Spiritual Vortex**' is said to be a cross-point between energy fields in the earth's grid system and intersecting ley lines that carry supernatural magnetic energies. We rarely see them with the naked eye but we can capture their energy effects in photos, and we can feel and sense their vibrational energies coming through.*

***Foco Tonal (Hunab Ku)** is one of these portal-vortex places where pockets of concentrated energies can be harnessed by certain individuals or by mass meditations, to cause **powerful surges of energies** from other planes of Celestial Activity. People standing inside this white brick circular sanctuary-shrine may feel a range of different sensations, such as a feeling of serenity, peace and even healing properties and a feeling of oneness.*

Note:
Graca, Juan and Maratt** are lightworkers who do meditation and channeling at Foco Tonal for the benefit of love, peace, health, nature and animals, in the name of all humanity.* ***(See photos, PAGES 139-142)

Fast flowing 'golden energy-beams of light'

Luminous bright 'energy fields of flashing lights'

Flashing–beams of light

CASE STUDY 5:

Theresa Kaplan Amuso (Arizona USA)

A portal-vortex is like an 'invisible space of energy' (invisible power spots) that can travel between worlds of different dimensions.
Gaia has many of these 'power spots' but have been systematically suppressed over hundreds of years.
*These **'power spot areas'** have on-going Energy Transfers like a busy airport, and anyone standing in such areas with a camera would be able to capture these fast moving energies.*

Very rare does Theresa get perfectly round orbs

I believe Theresa's house property, is built on a power spot which has on going activity all day long. ***The electro-magnetic energies in this area are travelling very fast, because it is 'very rare' to see perfectly round orbs that are at a stand-still.***
Theresa is a former field investigator and a former lightworker, and she says that the activity on her property opened up in 2011. This area looks like a busy cosmic traffic control area with fast moving electro-magnetic ***beams of light heading off in all directions****.* ***(See photos, PAGES 144-147)***

Please note:
*If we compare the two power spots of '**Hunab Ku in Mexico**' and '**Theresa's property in Arizona**', we see two 'different types' of electro-magnetic activity. Hunab Ku has bright light-bulbs and flashes of golden beams which often come during meditations or on asking them to come, and there seems to be a conscious connection with the people inside this round white-brick shrine. Whereas Theresa's property shows a different type of fast moving energy that seems to be constantly there 24/7 and constantly forming fast moving beams in 'every direction'.* ***Could they be 'different types' of power spots with different functions and missions, because they may be coming from different regions of the cosmos???***

*Electro-magnetic 'fast moving beams of light' moving in **'all'** directions.*

*Electro-magnetic 'fast moving beams of light' moving in **all** directions*

CASE STUDY 6:

Marion Atehsa Cyrus (Germany)

Marion loves to photograph flowers, plants, animals, forests, parks, gardens and mother nature in all her glory. She opens up to the consciousness of these energies and she captures an array of ***many forms of different types of orbs****, in many different shapes, forms and colours. Some orbs that she captures are perfectly round at a stand-still, and some are moving and brightly hovering around the trees, plants and flowers.*

Marion also captures a great variety of ***shape-shifting light energies that are in 'transformations' and some areas that look like 'spiral or circular vortex openings'.***
(See photos, PAGES 149-152)

Orb light energies and nature

Orbs and nature

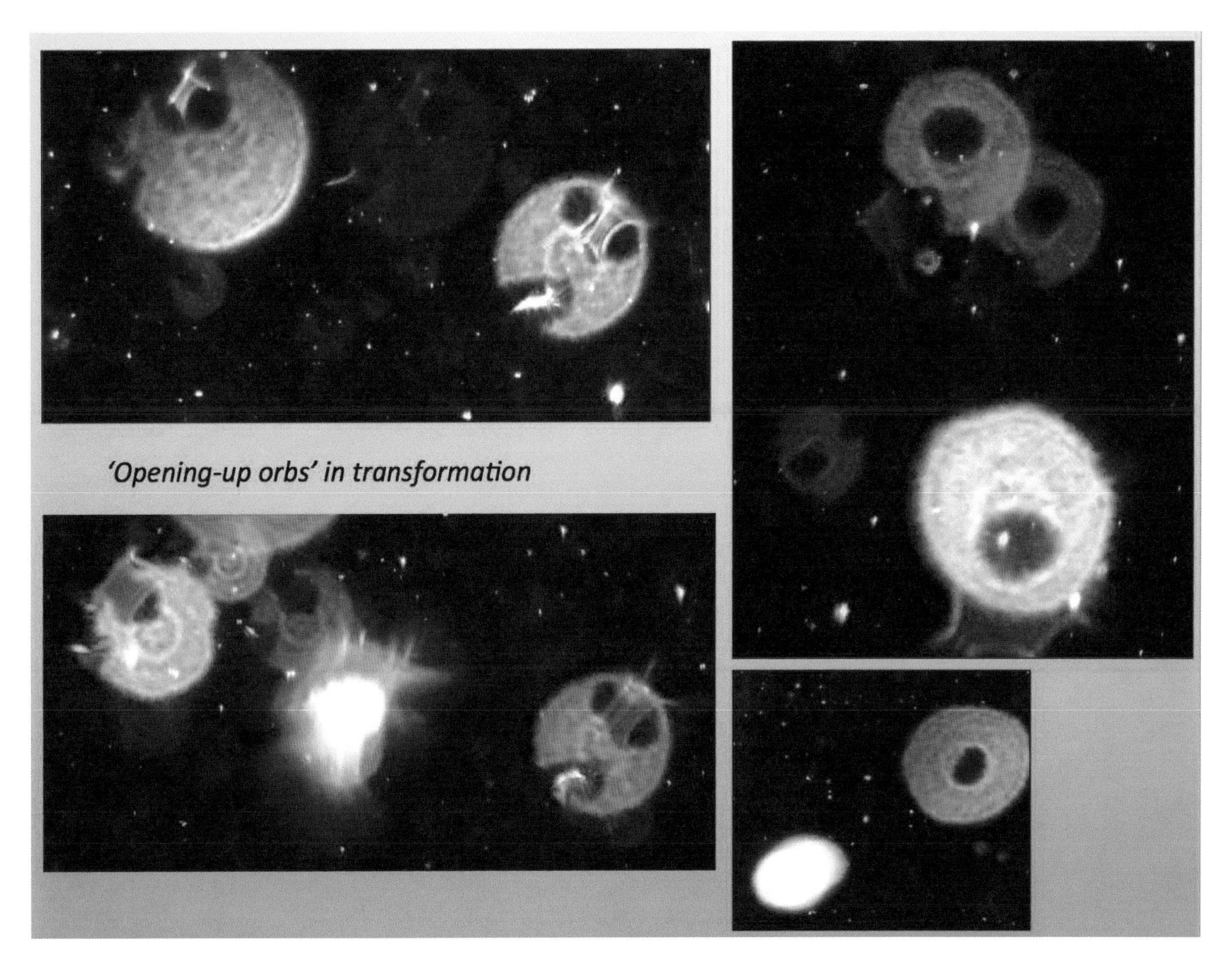
'Opening-up orbs' in transformation

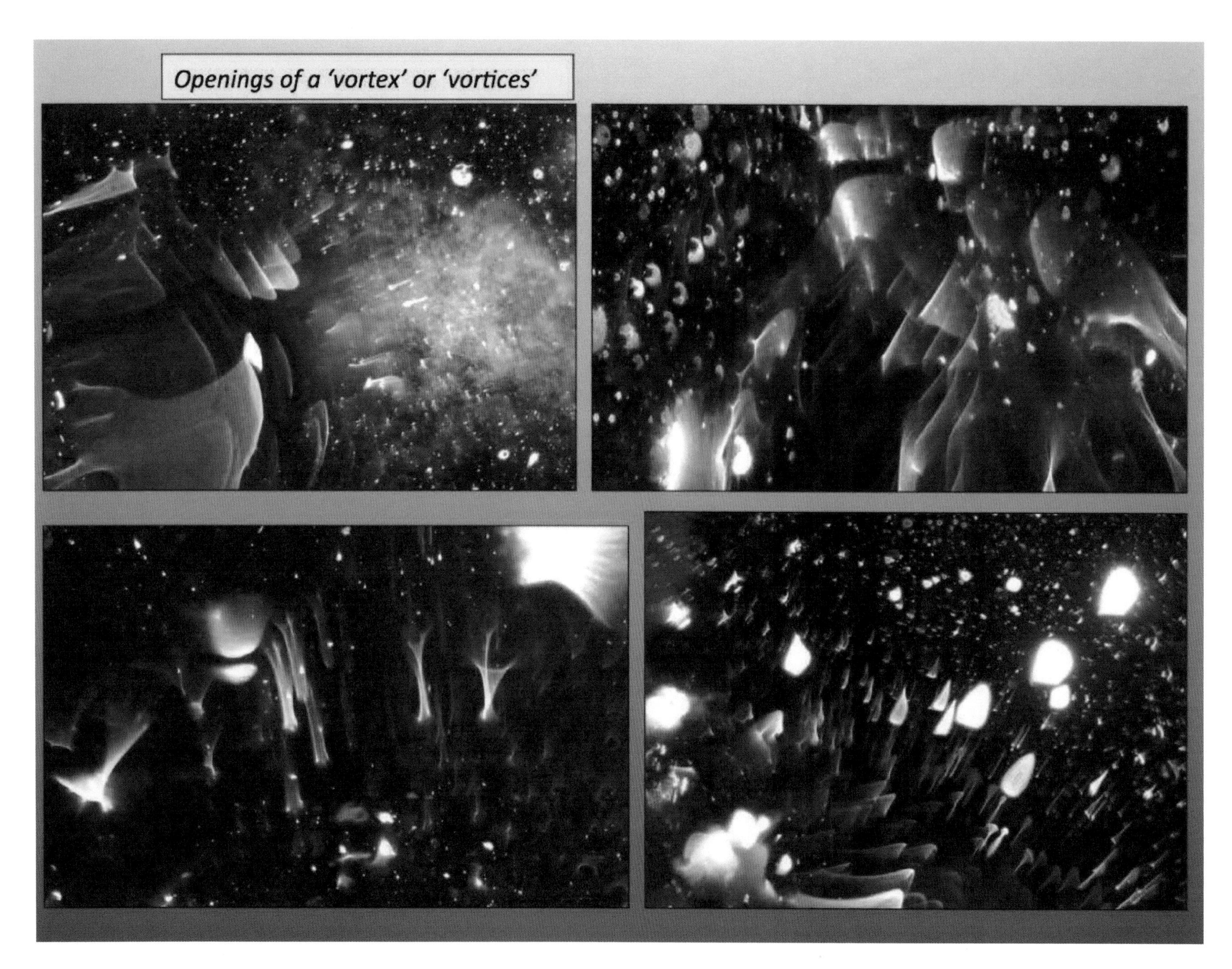
Openings of a 'vortex' or 'vortices'

CASE STUDY 7:

Andrea Corsick (USA)

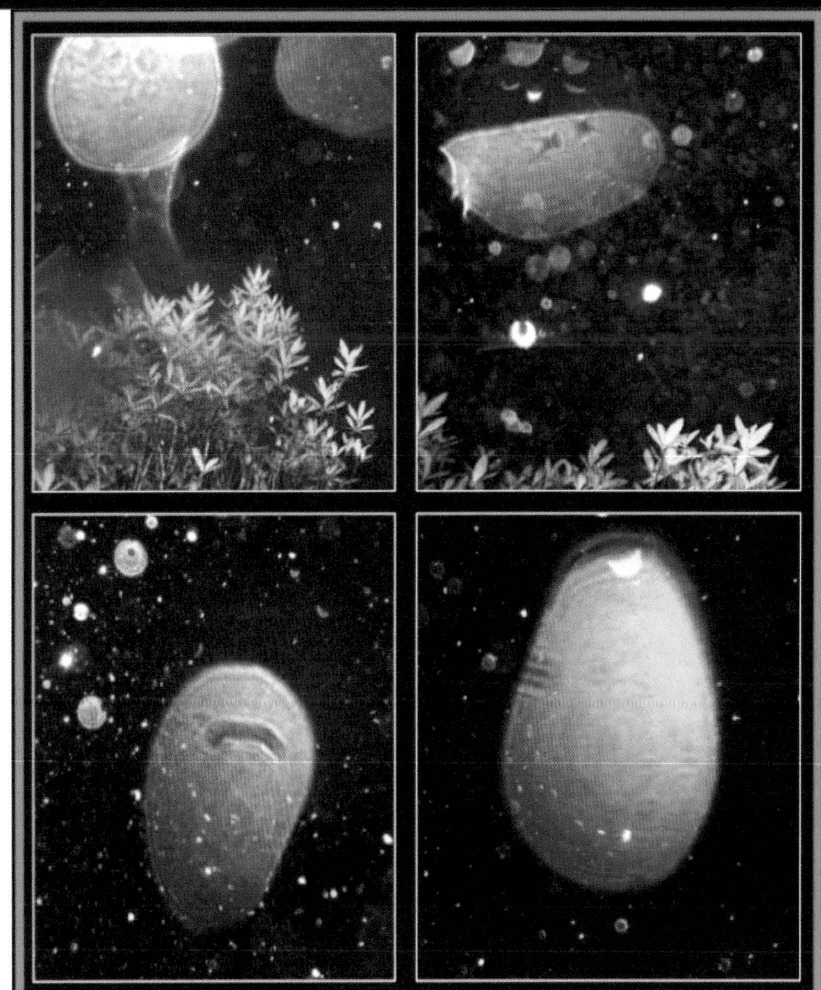

*Andrea's photos are an impressive display of the Orb Phenomena. She has a collectable variety and agglomeration of diverse energy fields which can be found in the Rhododendron Forest where she captures most of her photos. This forest seems to have pockets of concentrated energy fields that often show **formations of portal-vortex areas**. (A moving vortex often has angular, linear and circular momentums of energy, like a swirling mass or rotating centres that can create unusual forms of spirals or columns, and we can see this in many of her photos).*

*There are so many different variations in this area, and I believe that **Andrea is capturing orbs and energies from many different regions of the cosmos**, which may explain the many varieties and diversities of orbs that she is capturing..* ***(See photos, PAGES 154-160)***
Please note:
*(If we compare **Marion Atehsa Cyrus's photos** from Germany, to **Andrea's photos** in California, we can see many similarities of many similar type orbs).*

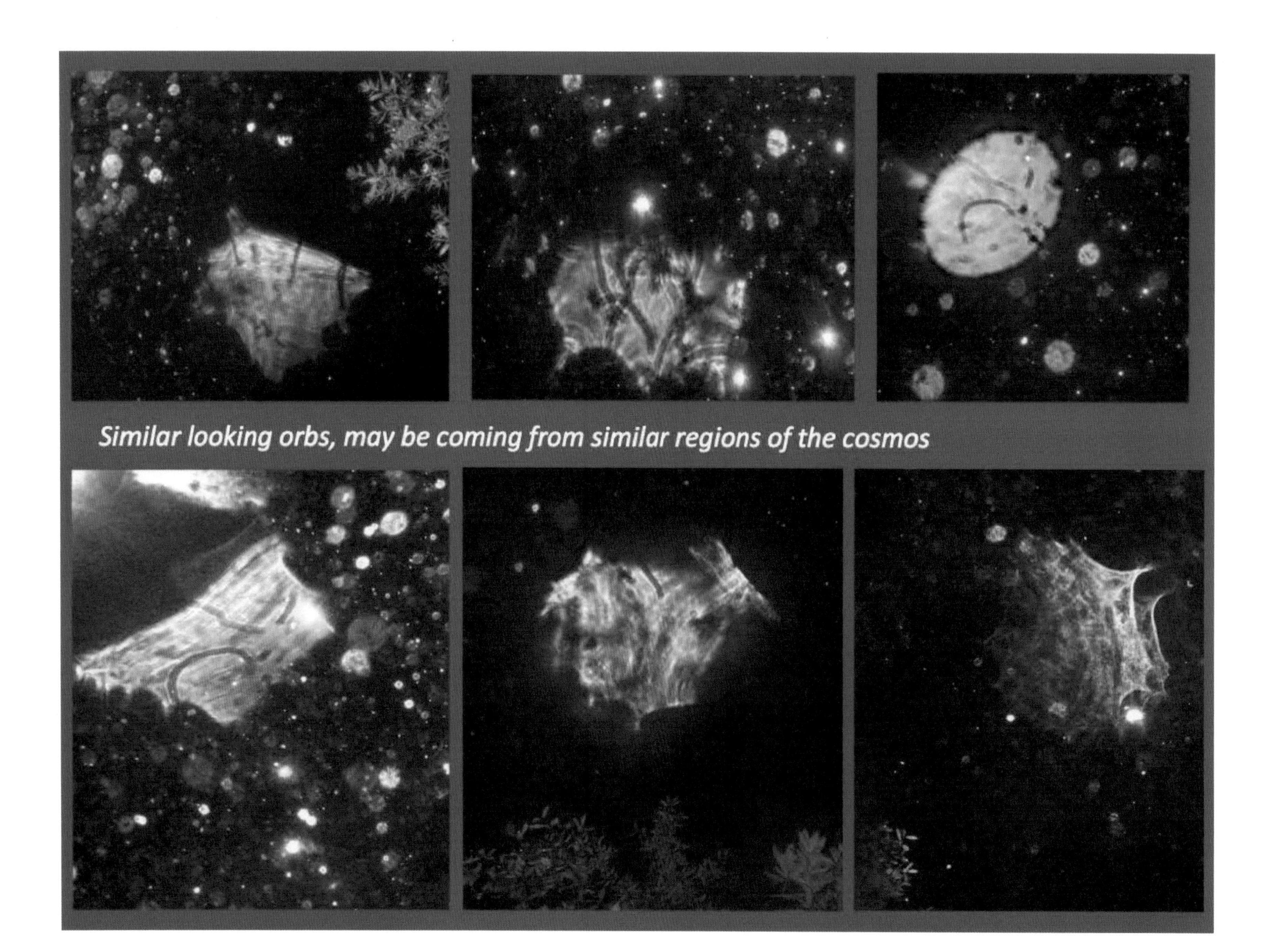
Similar looking orbs, may be coming from similar regions of the cosmos

Similar looking orbs, may be coming from similar regions of the cosmos

Similar looking orbs, may be coming from similar regions of the cosmos

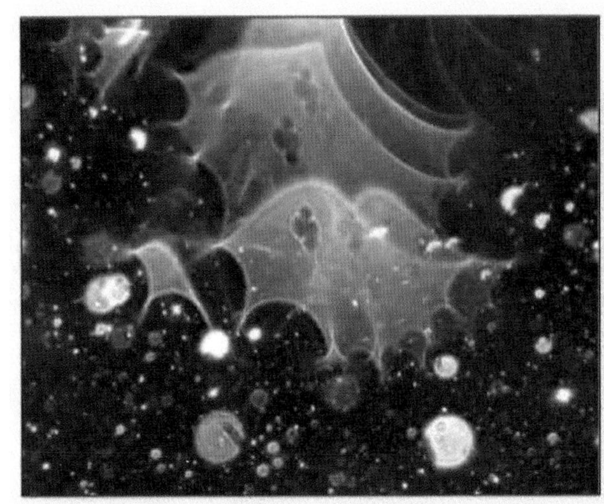
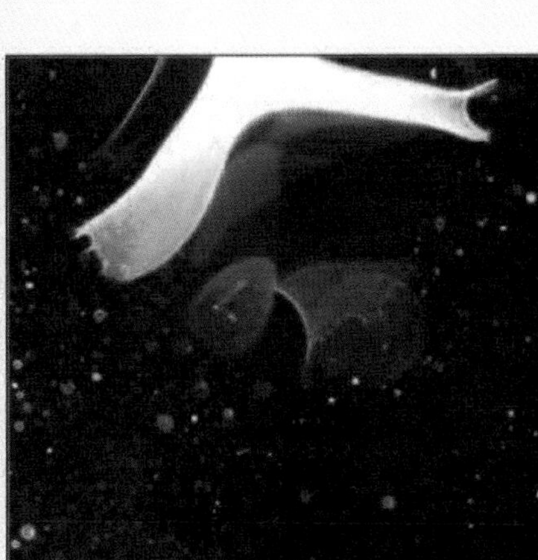
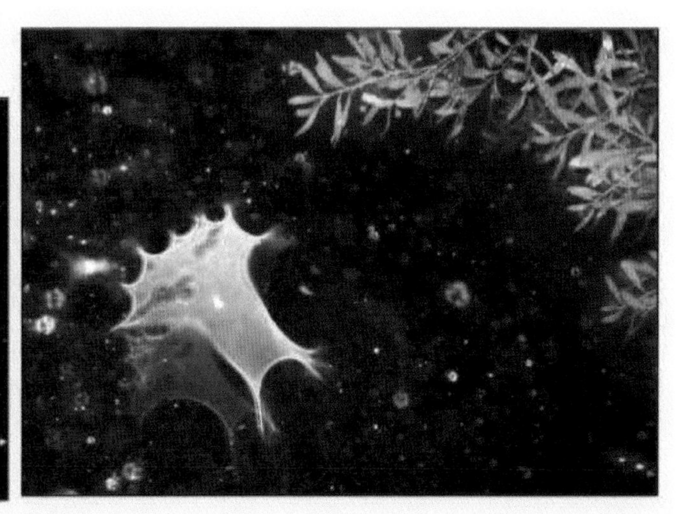

Many different variations

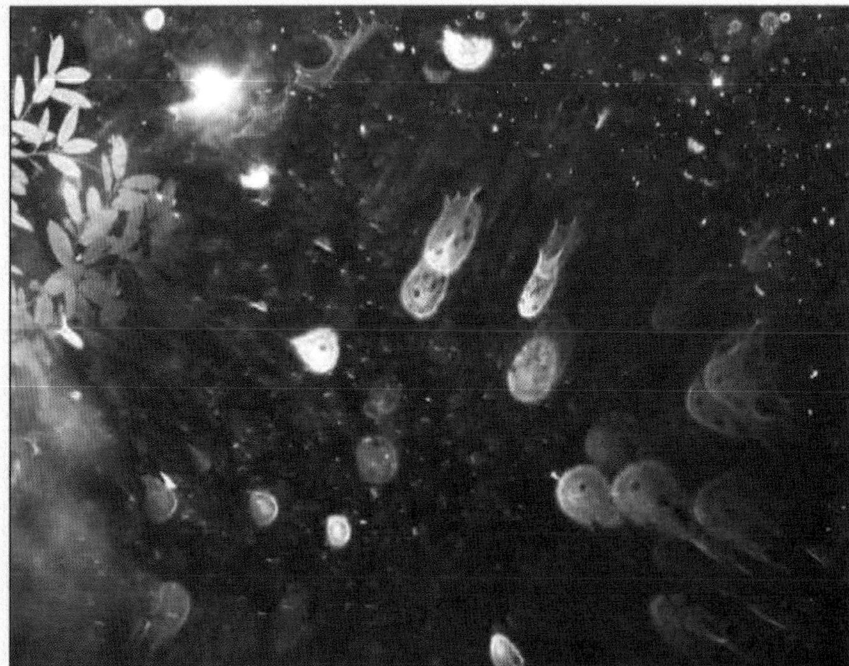

Openings of a 'vortex' or 'vortices' displaying
unusual forms of spirals, veils and columns

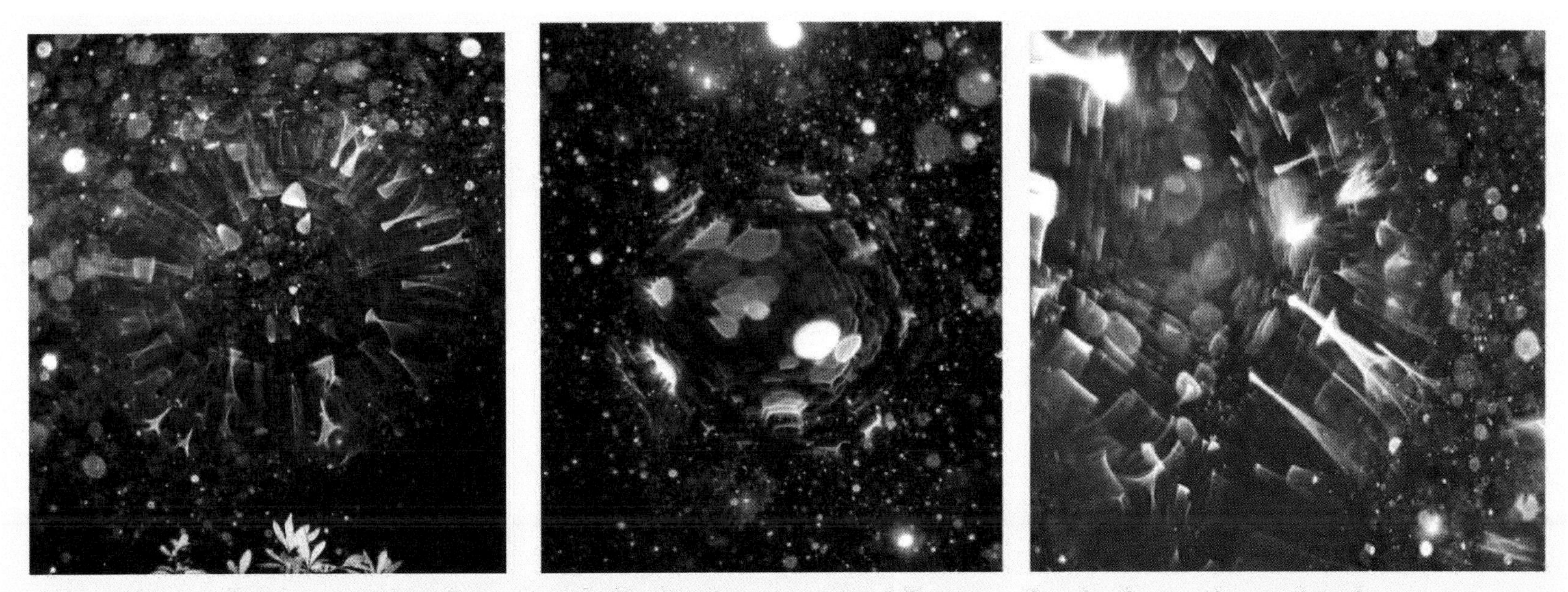

Openings of a 'vortex' or 'vortices' displaying unusual forms of spirals, veils and columns

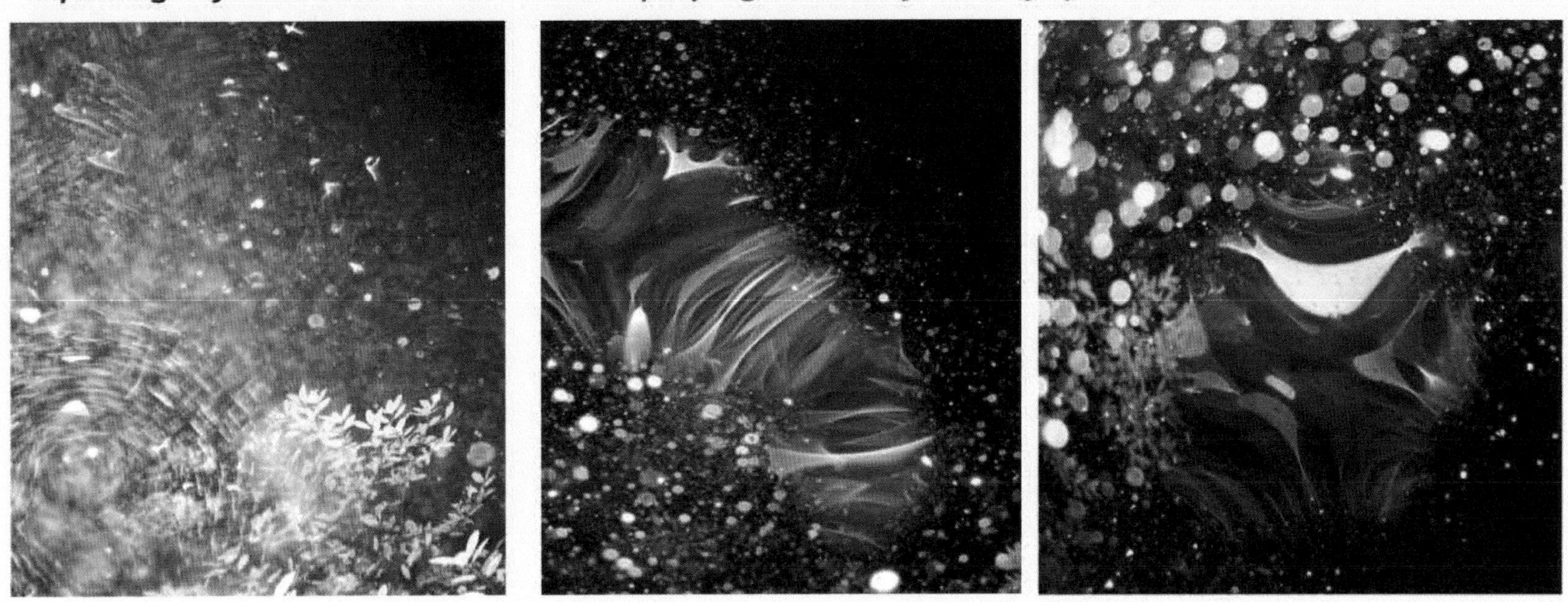

'Orb energies' create
beautiful wild colour
explosions of art
above the trees

Part 15
Metaphysics vs. Physics vs. Scientific Research

Some of the most shattering 'truths' of our reality and the unknown universe haven't even been discovered yet. And the borderline between **Physics** and **Metaphysics** is defined by whether a hypothesis or theory is 'experimentally testable'. If it is not testable and we cannot humanly understand it, or science sometimes just runs out of answers, then most often it is just swept under a carpet and kept out of sight!

All scientific studies should be completely 'objective'. They should not be influenced by religious opinions, political opinions, biases, personal and emotional opinions or manipulated and tampered results. The researcher should strive to be objective with his/her observations that may either help support or help refute a scientific hypothesis. And in general, raw data is considered evidence only if it has been interpreted in a way that meets researchable and testable criteria for scientific knowledge. But despite the fact that science is continuously advancing with new technologies and new science research, we do not know everything about our own world and the Earth we live in, not to say the least for other alternative realities or other worlds.

For a 'Hypothesis' to become a 'Theory', it is crucial that it is consistent, empirically testable, correctable, improvable, progressive, and can lead to reproducible results so that others can double-check it. It must be able to resolve current paradoxes and create a *new paradigm and new questions to work on.* And above all it must be parsimonious in the adoption of the simplest assumption in the formation of a theory or in the simple interpretation of the data acquired.

The scientific method of procedure of acquiring knowledge has characterized the development of science since the 17th century. ***This involves careful observation, experiment, testing and applying rigorous skepticism about what is observed and formulating hypotheses based on experimental findings. (This is what I believe that I have been doing over the ten years of my own personal research, observing, studying, testing and hypothesizing on the mystical and very misunderstood 'ORB PHENOMENA').***
'All' pioneering images of my research work are genuine and authentic. I do not in any way whatsoever tamper with, or falsely re-construct the images of the sentient beings that I capture. My 'MFT Method' is a simple straightforward

procedure that can consistently, be replicated over and over again, that cannot be dismissed as fake. **I am capturing images of 'life beyond our reality'!**

Are they Aliens from interplanetary, interstellar or intergalactic regions, or Angelic beings from the Heavenly realms? I cannot answer this because I do not know!

But I *do* know that by discovering my cutting-edge MFT Method, I am able to visually focus into these hologram-type capsules, or projections of the 'sentient beings' that use orbs as astral travel.

> Note:
>
> The best way that I can describe what I am capturing, is to compare an ORB as a 'TELEVISION SET'. For example, if we are watching people on the television screen, we are watching '*AN IMAGE*' of them, which is being projected on the screen. From a 3-D reality of them, we are seeing a ***2-D IMAGE*** of them, projected on the television screen, because one of the three co-ordinates is *eliminated* in order for us to see them on a 2D screen.
>
> **However, Orbs are not 3 Dimensional, they are from MULTIPLE DIMENSIONS**. We are seeing an image of them that is being projected to us from inside the '***Orb Screen***' that has been captured by a 'camera'. Due to their very complex co-ordinates, I believe that they are showing up as 'Simple Holographic-type Images' because some of their co-ordinates have been '*eliminated by the camera*'. I believe that when our technology improves, WE WILL BE SEEING THEM MUCH CLEARER AS 'HOLOGRAPHIC IMAGES INSIDE THE ORB SCREENS'.

I have added a few pages here of ***Real and Authentic*** footage photos from my research files of ***real sentient beings,*** from which I have over 5,000 images of different sentient photos from different orbs.

These images are definitely ***not*** pareidolia! There is a definite system in orbs, and they follow specific patterns, rank levels and specific characteristics.

(See photos, PAGES 163-171)

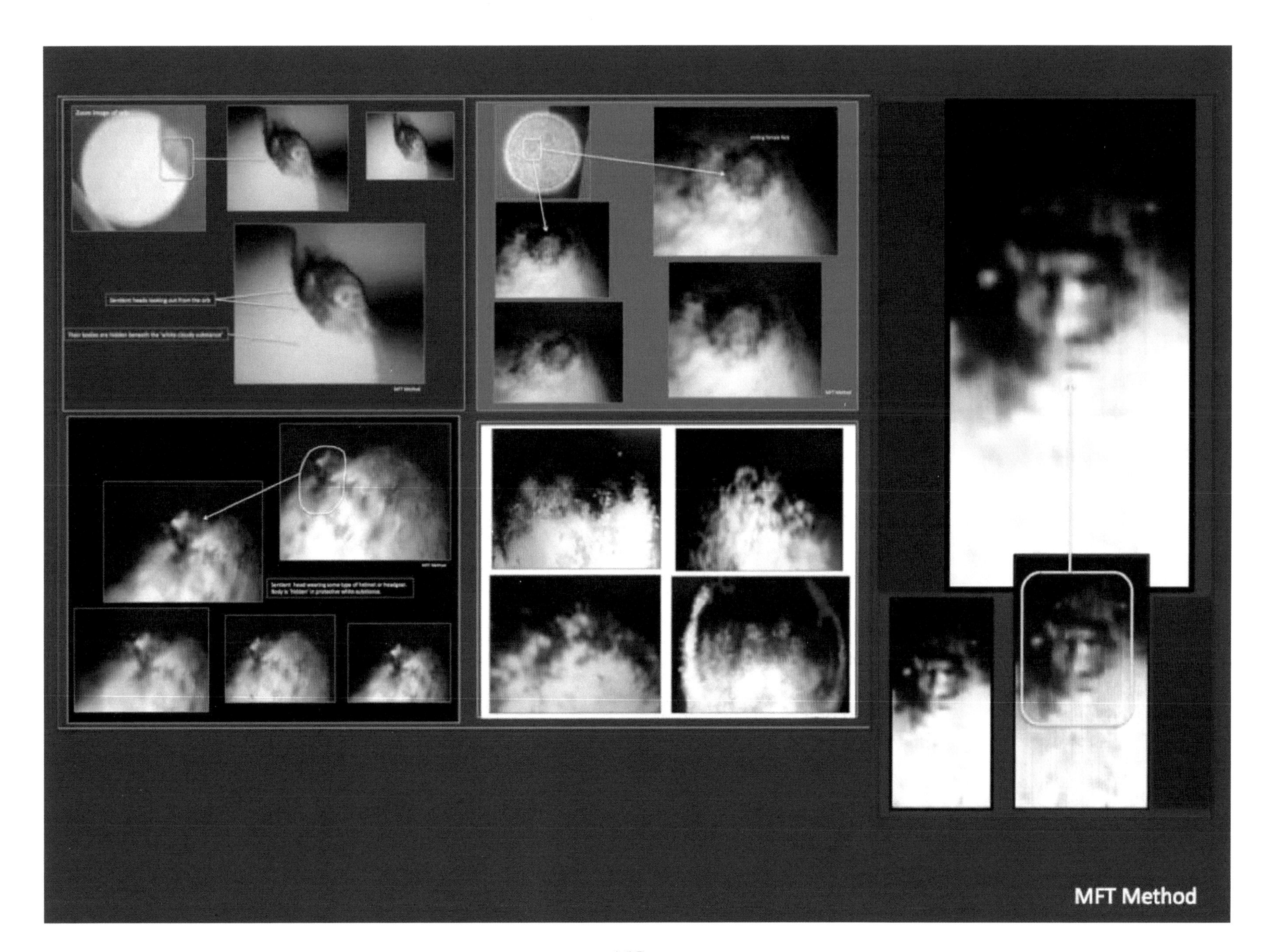
MFT Method

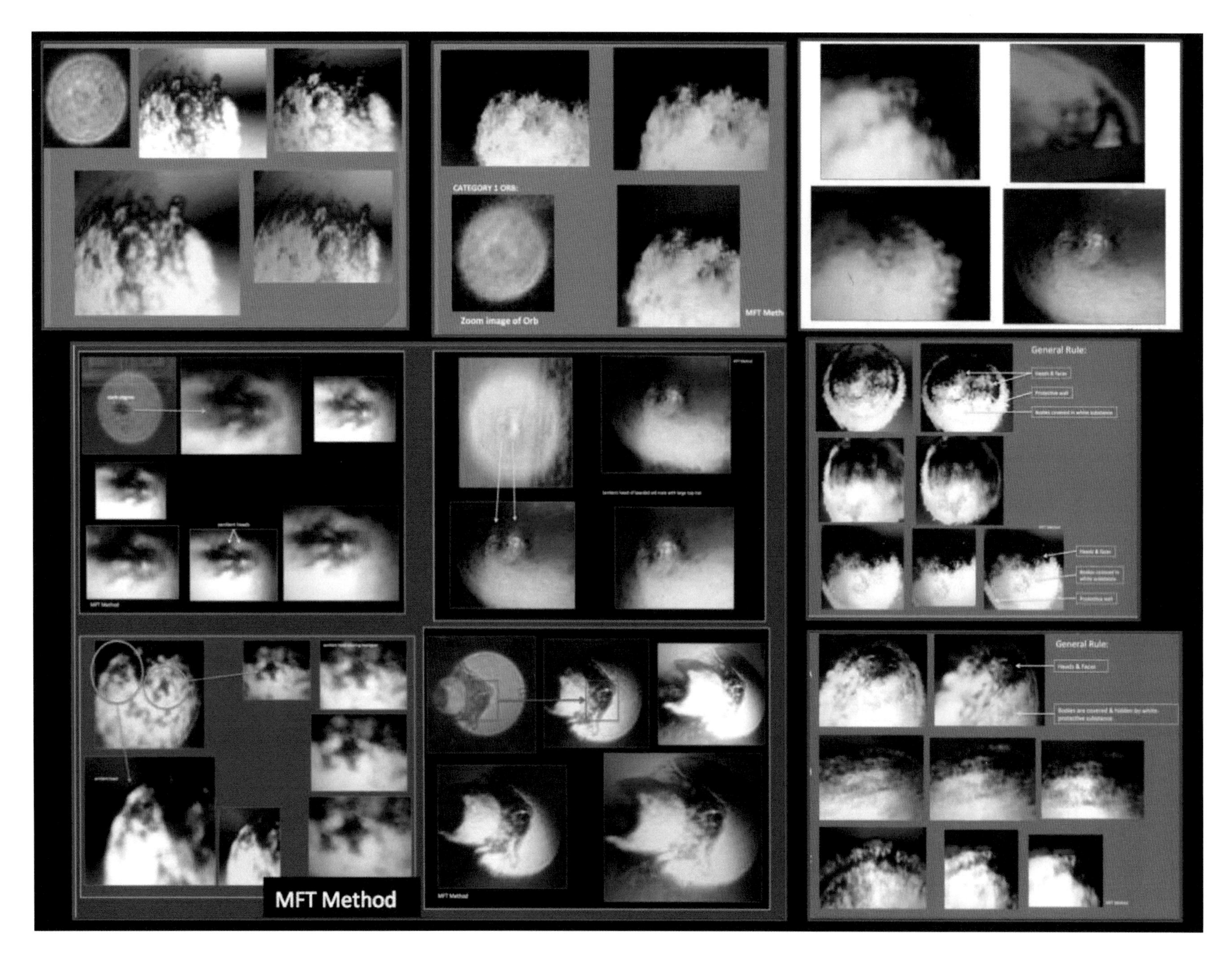
CATEGORY 1 ORB:
Zoom image of Orb
MFT Meth
General Rule:
General Rule:
MFT Method

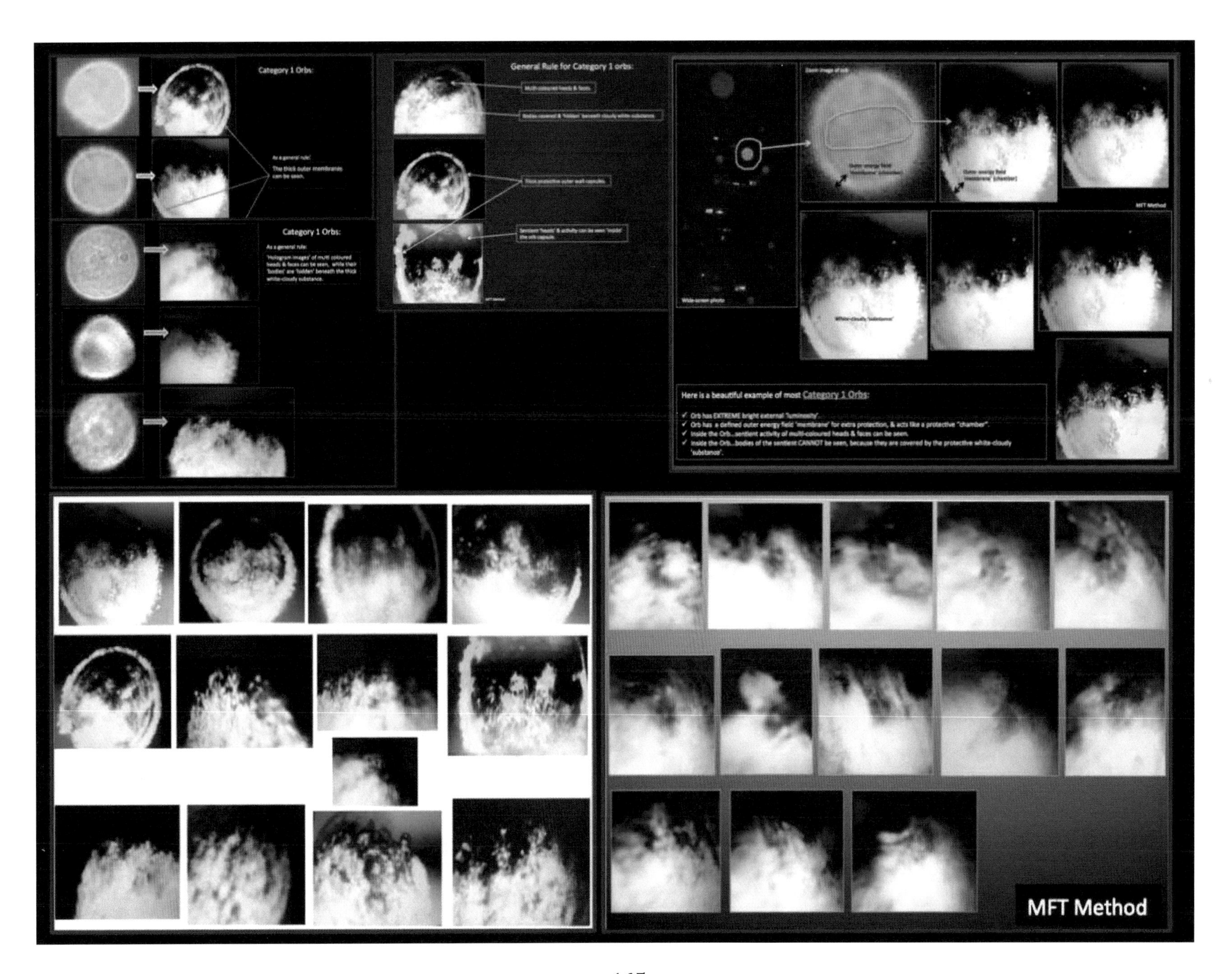

Category 1 Orbs:
As a general rule:
The thick outer membranes can be seen.
Category 1 Orbs:
As a general rule:
'Hologram images' of multi coloured heads & faces can be seen, while their 'bodies' are 'hidden' beneath the thick white-cloudy substance.
General Rule for Category 1 orbs:
Here is a beautiful example of most Category 1 Orbs:
✓ Orb has EXTREME bright external 'luminosity'.
✓ Orb has a defined outer energy field 'membrane' for extra protection, & acts like a protective "chamber".
✓ Inside the Orb...sentient activity of multi-coloured heads & faces can be seen.
✓ Inside the Orb...bodies of the sentient CANNOT be seen, because they are covered by the protective white-cloudy 'substance'.
MFT Method

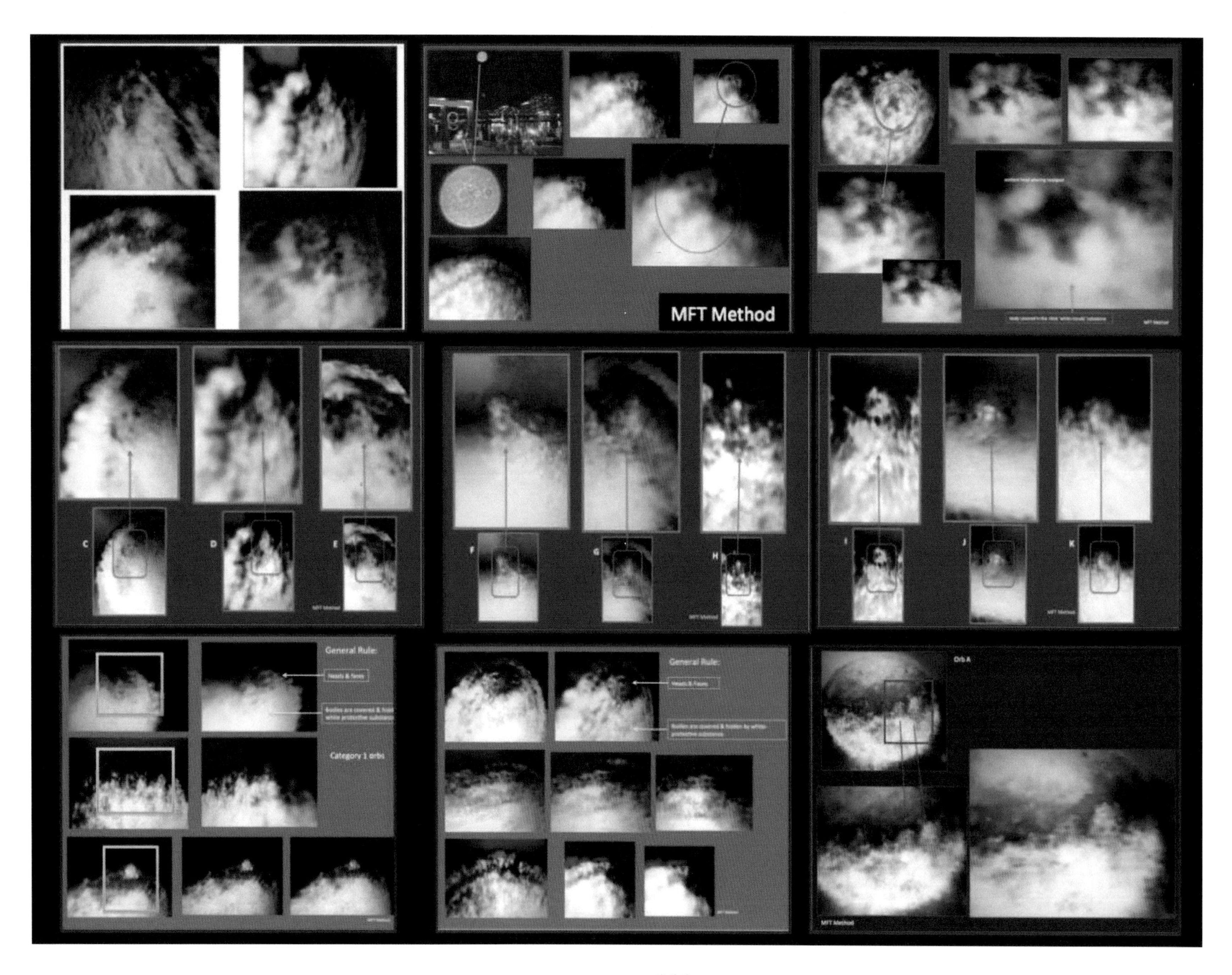
MFT Method
C
D
E
F
G
H
I
J
K
General Rule:
Category 1 orbs
General Rule:
Orb A

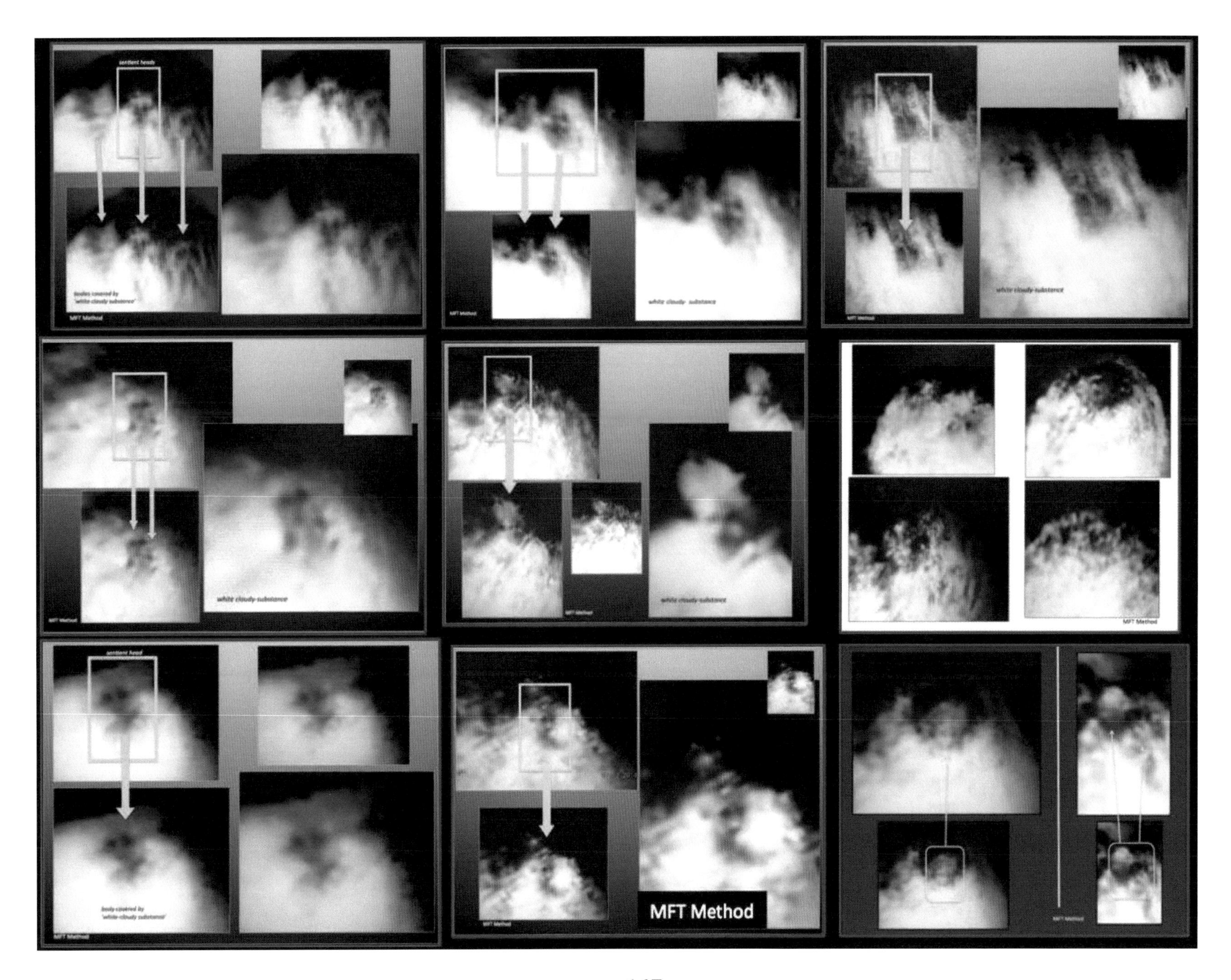
white cloudy-substance
MFT Method

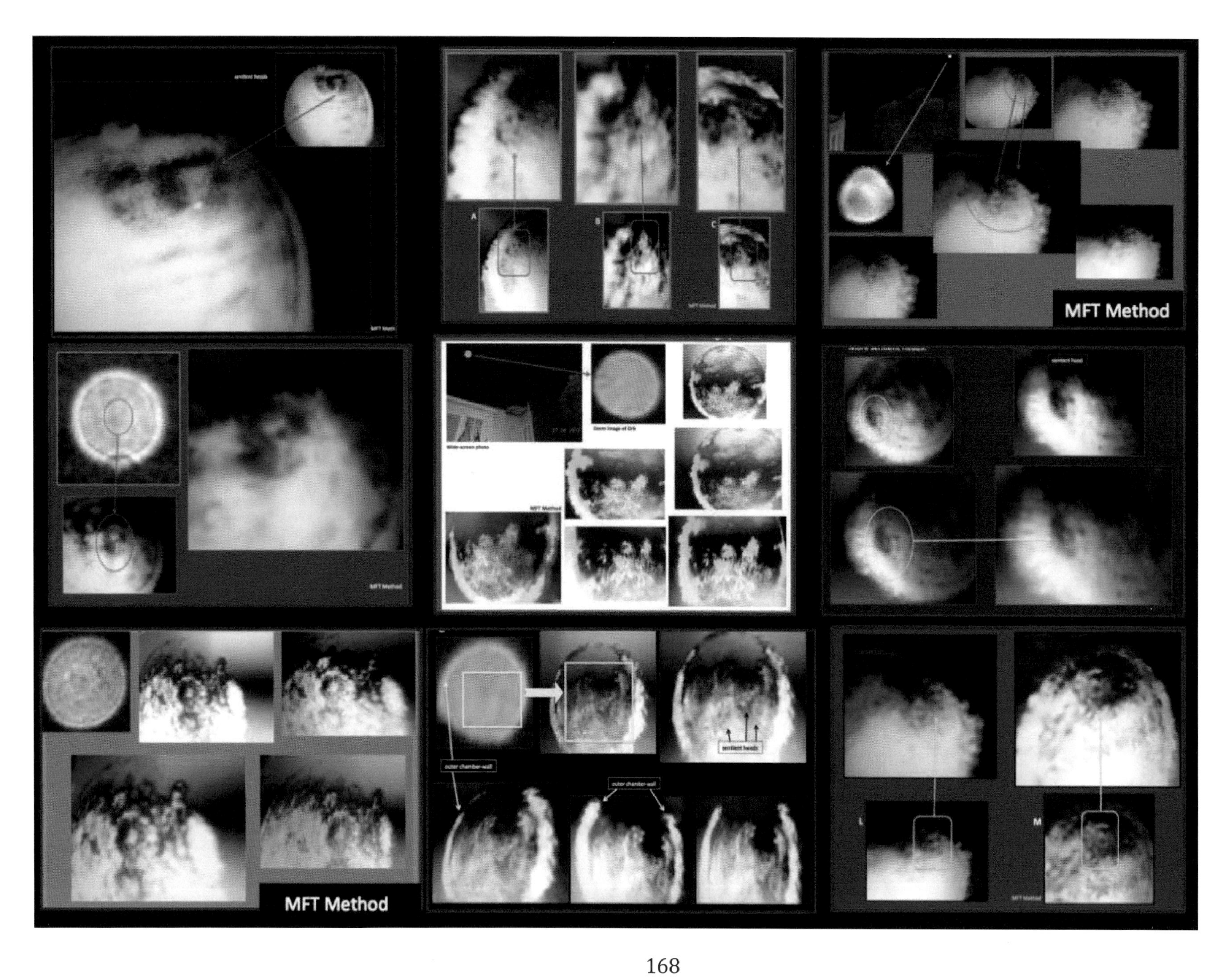
A
B
C
MFT Method
outer chamber-wall
sentient heads
outer chamber-wall
L
M
MFT Method

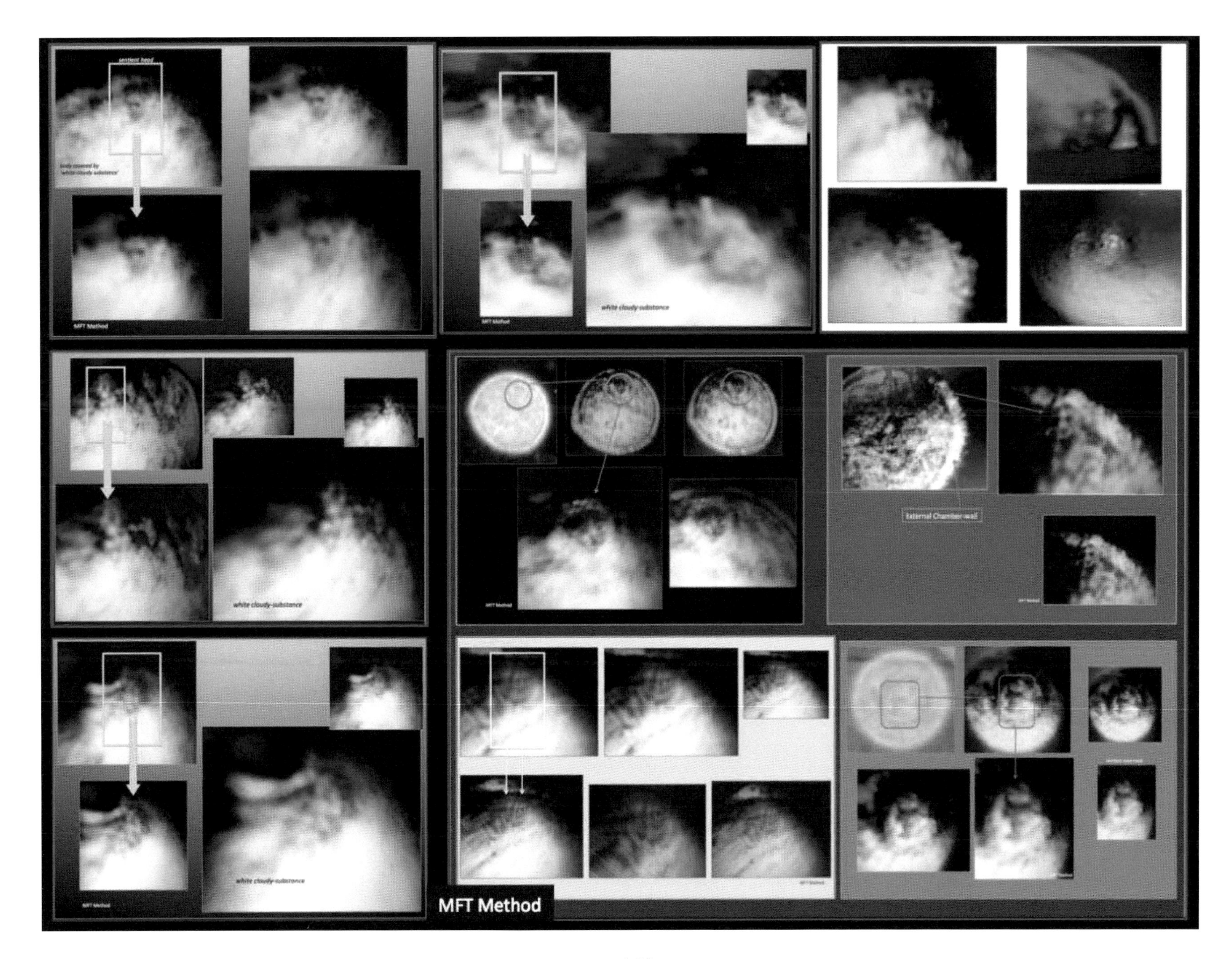

MFT Method
white cloudy-substance
white cloudy-substance
MFT Method
External Chamber-wall
white cloudy-substance
MFT Method

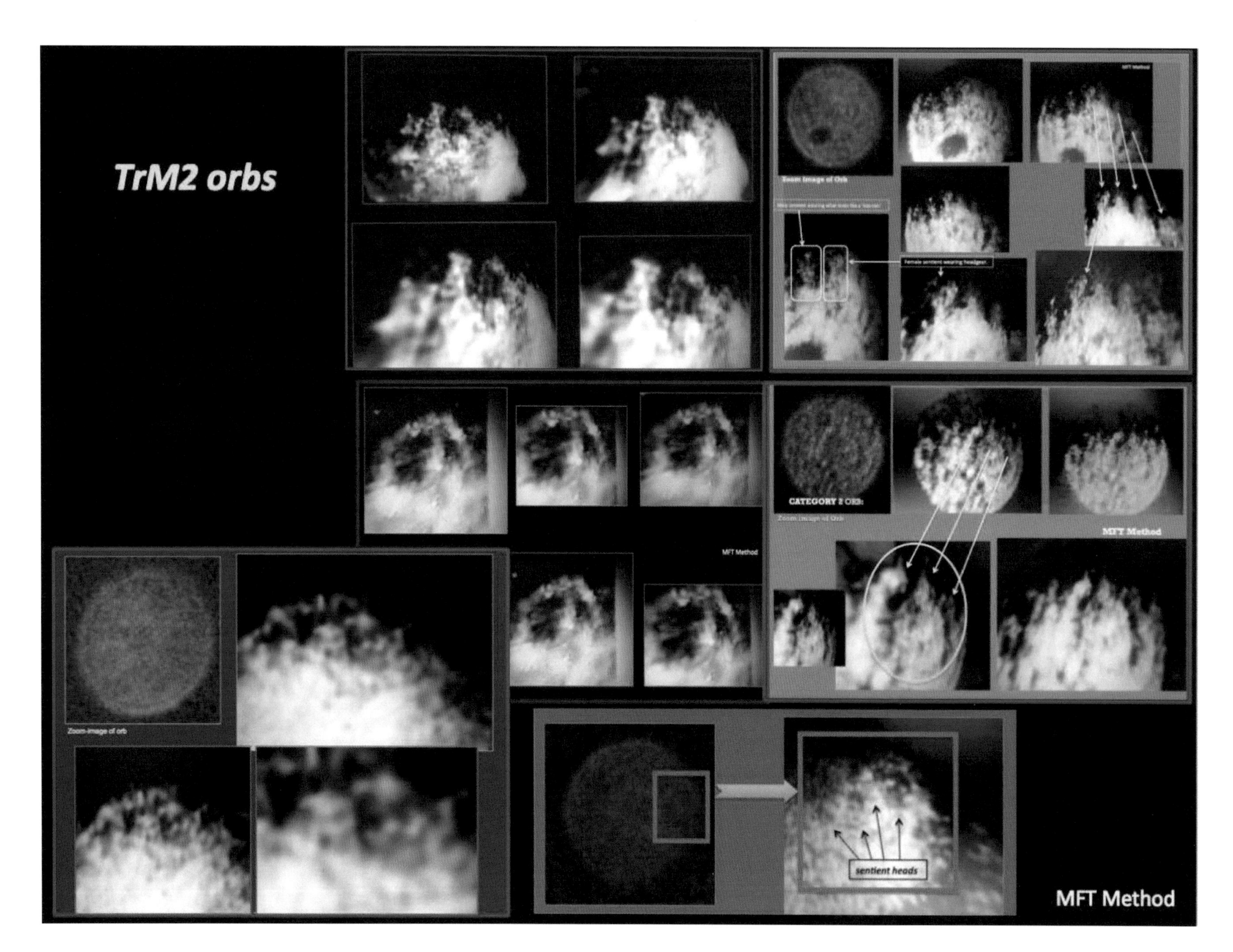
TrM2 orbs
MFT Method
CATEGORY 2 ORB:
MFT Method
Zoom-image of orb
sentient heads
MFT Method

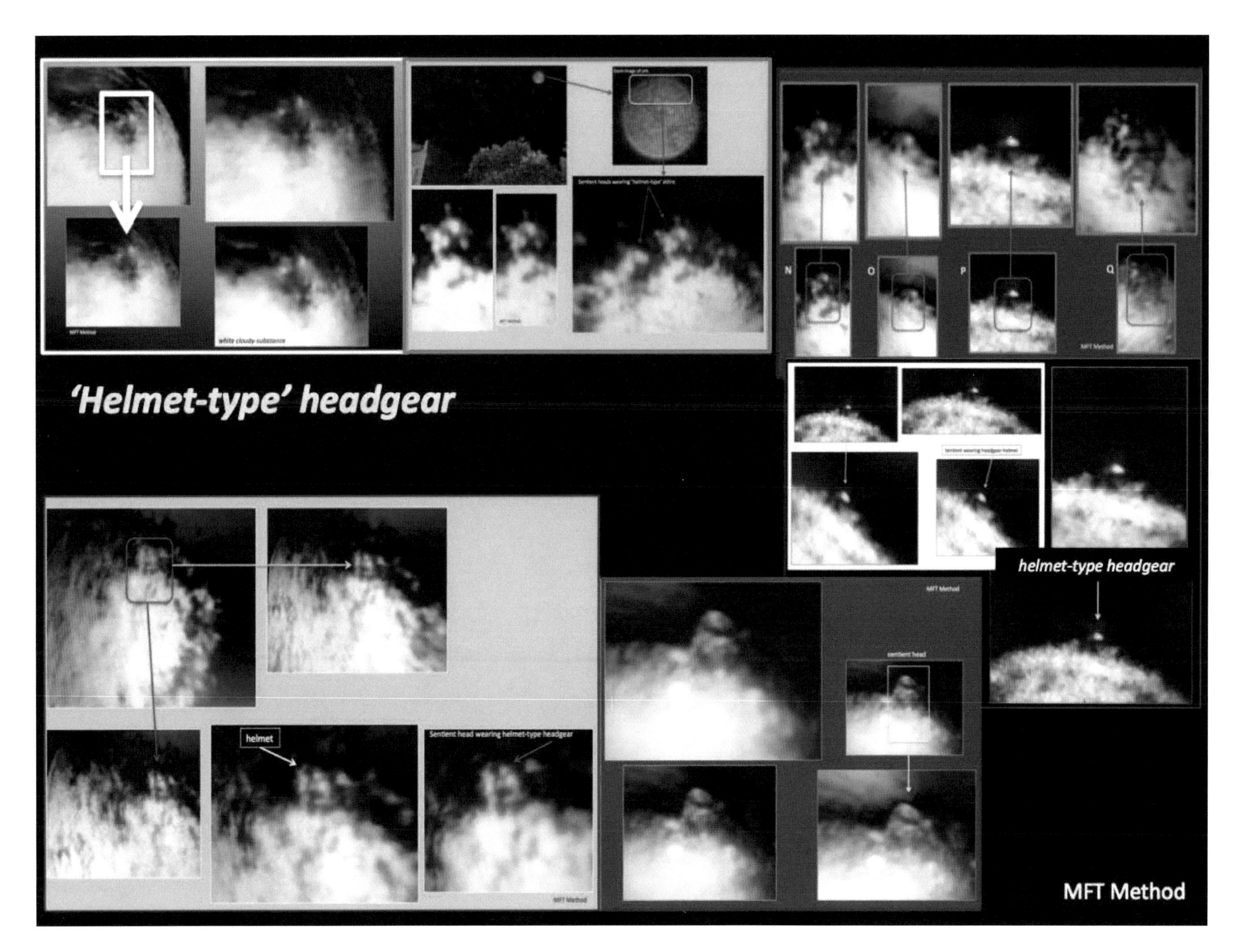
MFT Method
white cloudy-substance
Sentient heads wearing 'helmet-type' attire.
N
O
P
Q
MFT Method
'Helmet-type' headgear
helmet-type headgear
MFT Method
sentient head
helmet
Sentient head wearing helmet-type headgear
MFT Method
MFT Method

Part 16
My Research

Every Orb I have studied is always unique, with different colours, different faces, different scenes and different numbers of sentient images inside. Let's understand that 'orbs' are not the actual sentient beings themselves, but in fact, they are the travel modules, like a type of vehicle or cosmic rocket that encloses all the data and information inside. These globular or spheroid modules are made of materials unknown to us here on Earth.

Orbs are made from elements that can be thickly opaque, and yet have transparent qualities like looking through glass or water. Made from strong, durable weightless substances but at the same time they can be very flexible and very pliant. Whatever these mystical materials are, they can also penetrate and go through solid objects, they can travel at unbelievable speeds, and then at will they can suddenly immobilize themselves to zero motion and 'pose for us', so we can take photos of them.

I have added here, 13 more different orb examples of my research in more detail, so that you can see how orbs follow specific patterns and characteristics.
(See photos, PAGES 173-201)

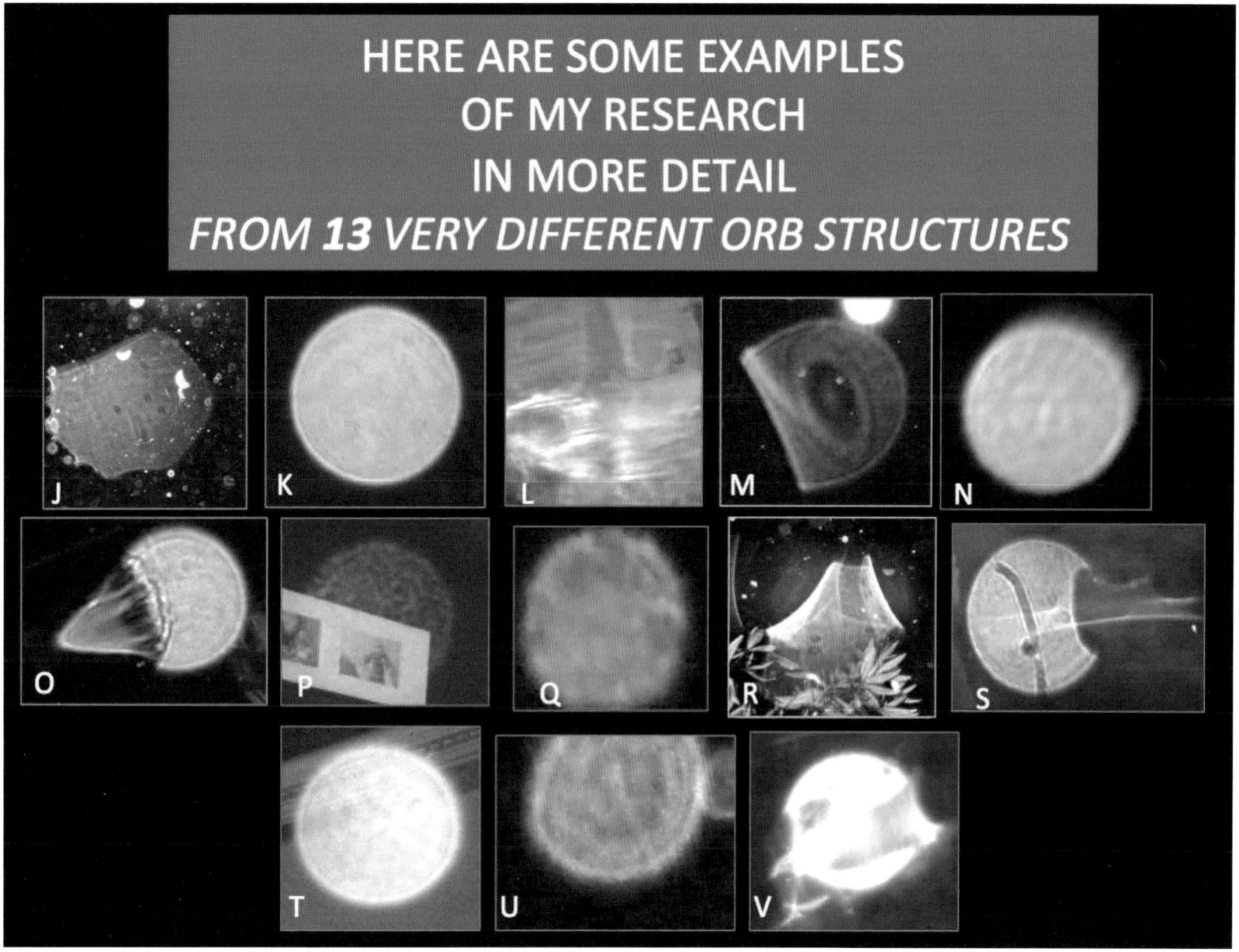

HERE ARE SOME EXAMPLES
OF MY RESEARCH
IN MORE DETAIL
*FROM **13** VERY DIFFERENT ORB STRUCTURES*

J K L M N

O P Q R S

T U V

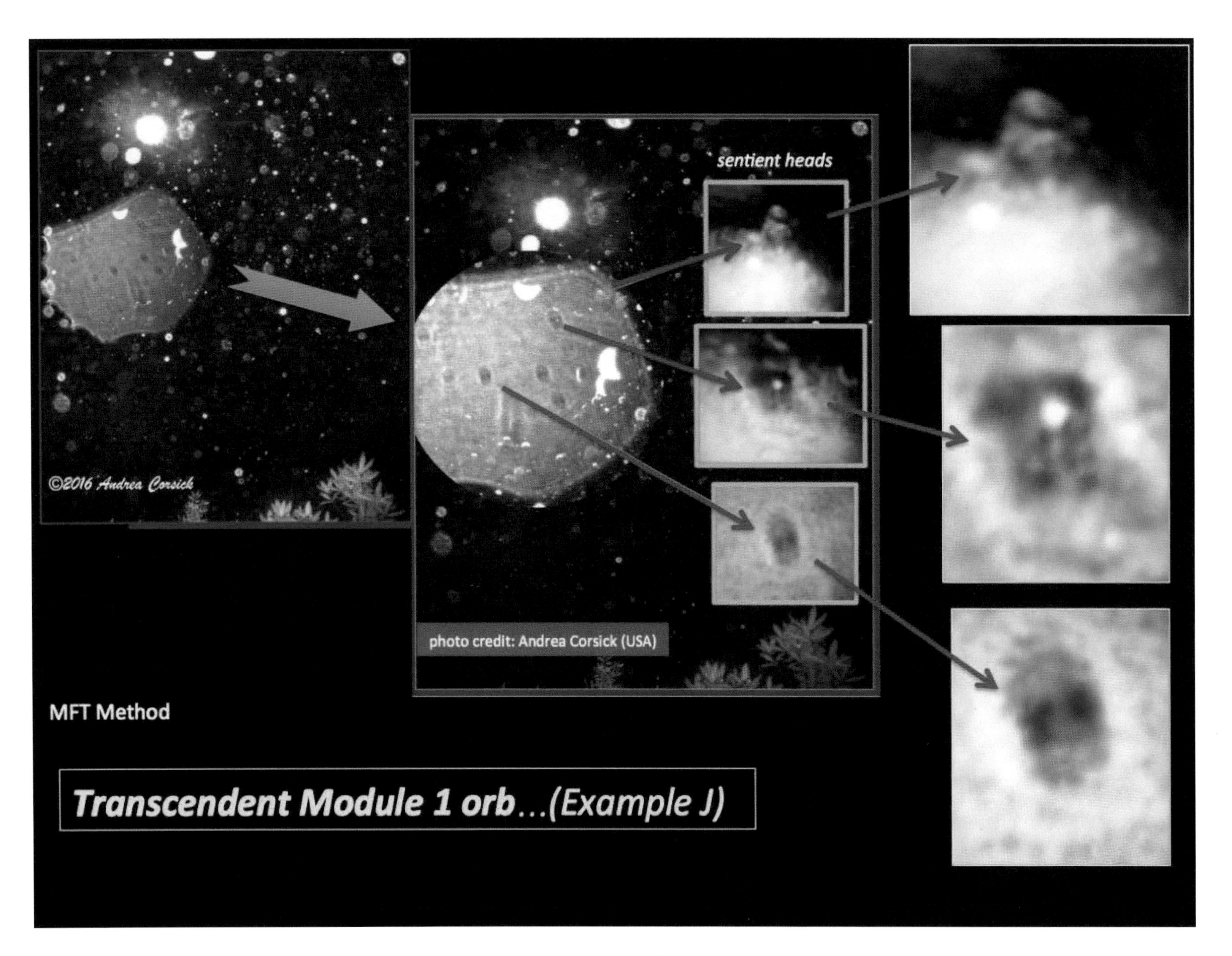
©2016 Andrea Corsick
sentient heads
photo credit: Andrea Corsick (USA)
MFT Method
Transcendent Module 1 orb...(Example J)

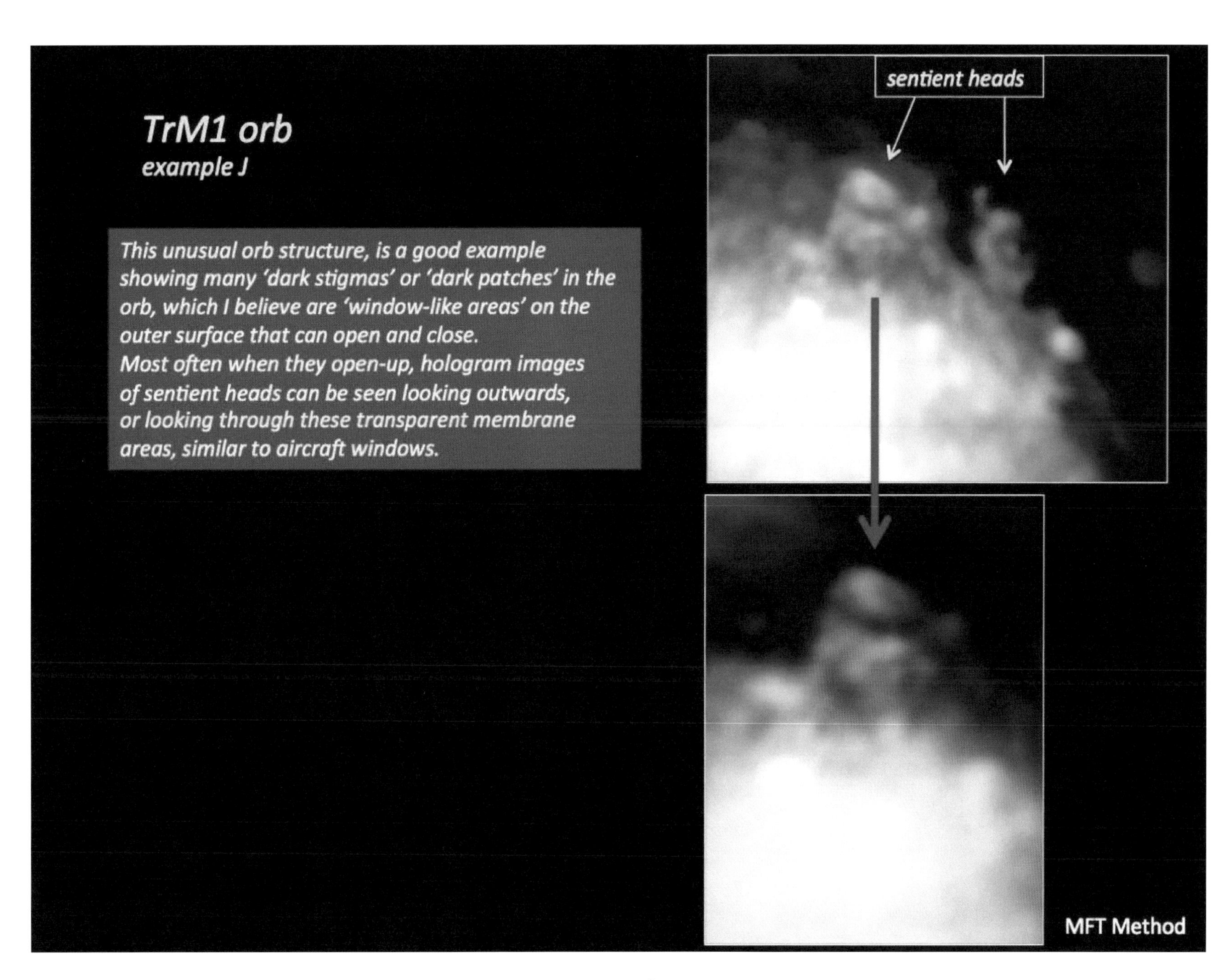
TrM1 orb
example J
This unusual orb structure, is a good example showing many 'dark stigmas' or 'dark patches' in the orb, which I believe are 'window-like areas' on the outer surface that can open and close.
Most often when they open-up, hologram images of sentient heads can be seen looking outwards, or looking through these transparent membrane areas, similar to aircraft windows.
sentient heads
MFT Method

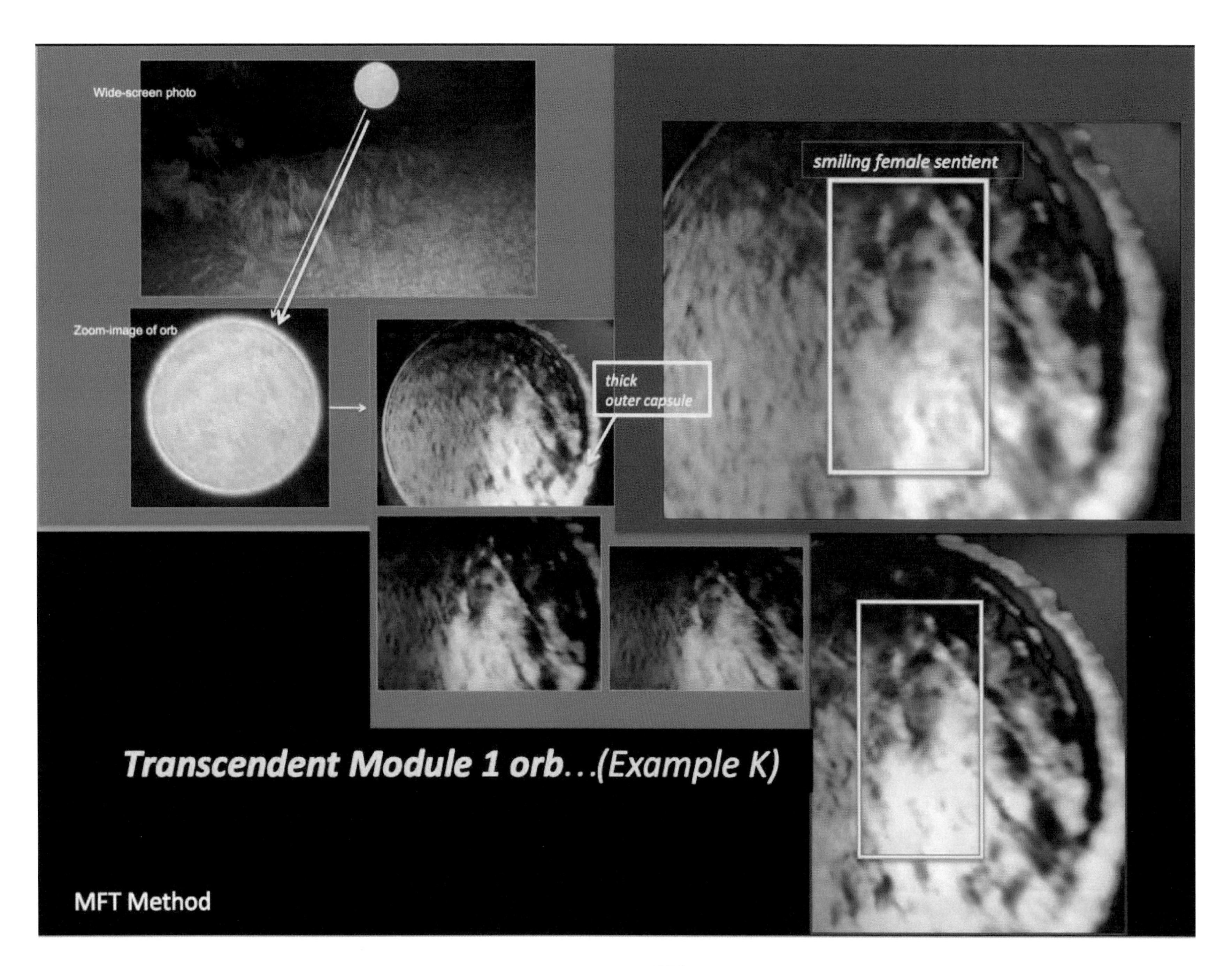
Wide-screen photo
Zoom-image of orb
thick
outer capsule
smiling female sentient
Transcendent Module 1 orb...(Example K)
MFT Method

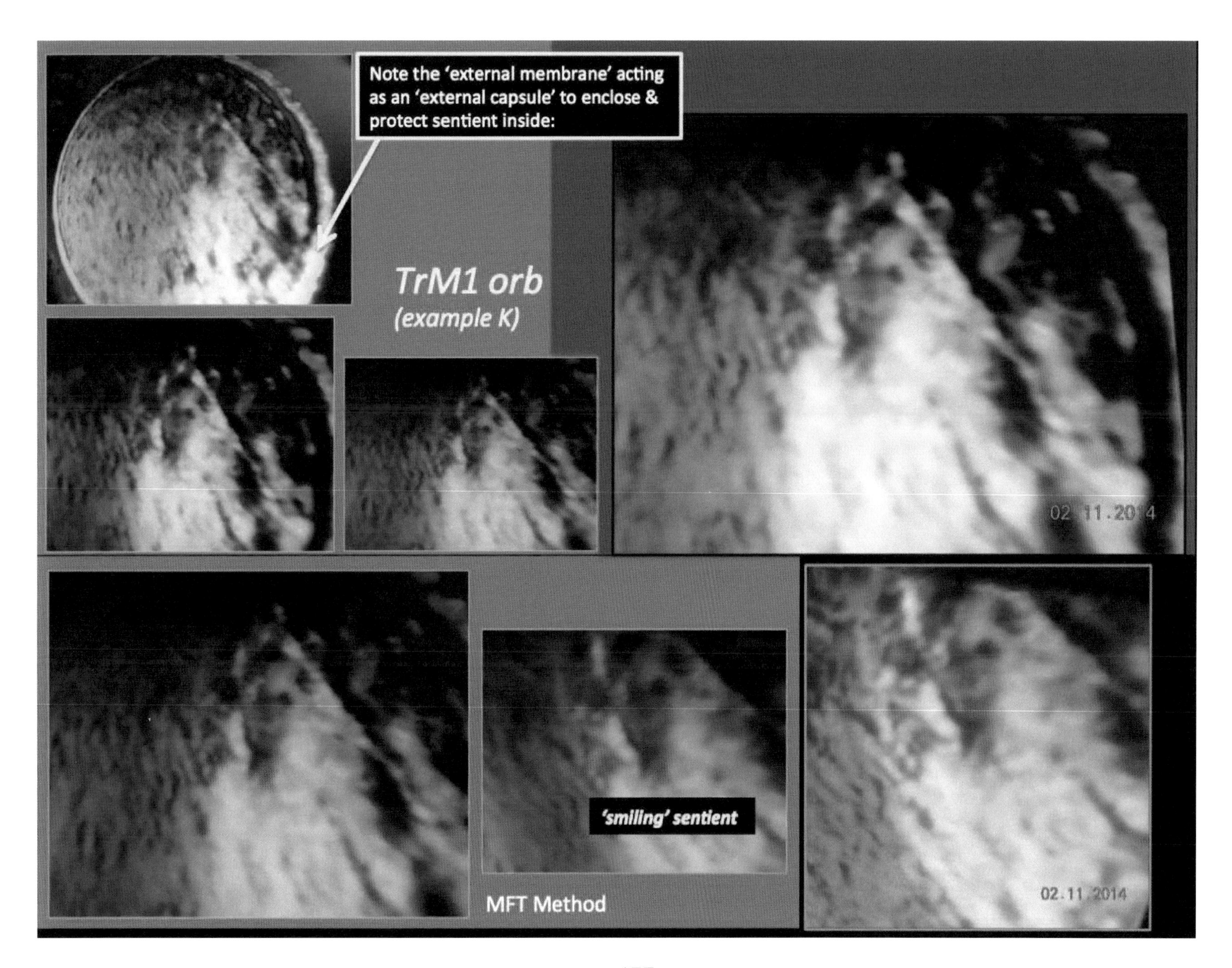
Note the 'external membrane' acting as an 'external capsule' to enclose & protect sentient inside:
TrM1 orb
(example K)
02.11.2014
'smiling' sentient
MFT Method
02.11.2014

***Transcendent Module 2 orb**...(Example L)*

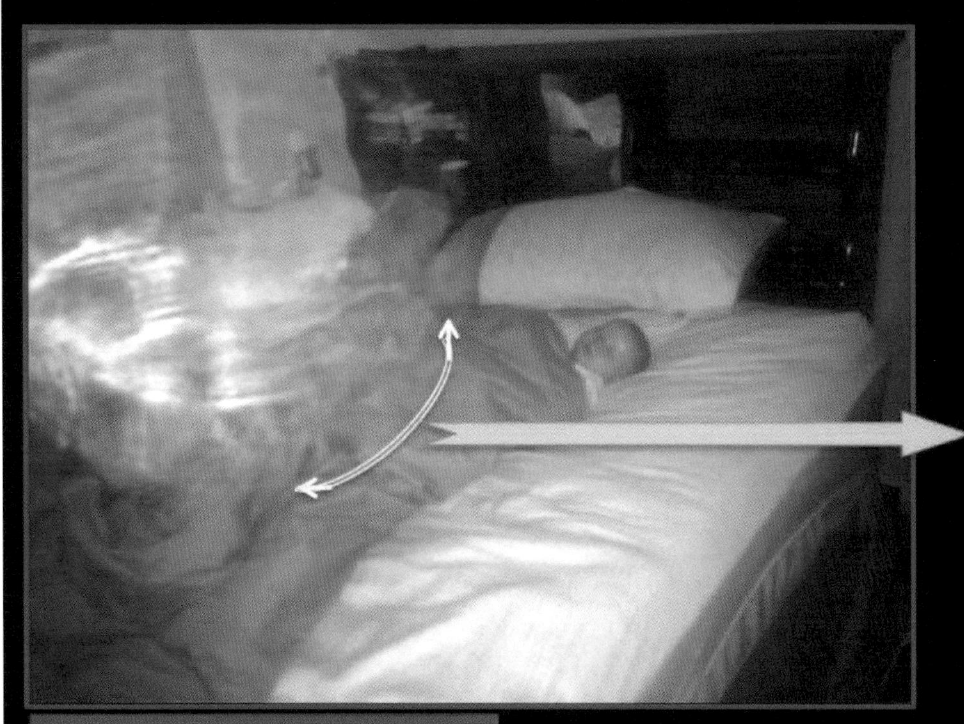

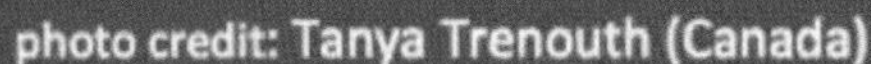

photo credit: Tanya Trenouth (Canada)

Quote from Tanya Trenouth:
"I took this photo of my son Gabriel sleeping in our bed. The room was semi-dark, curtains were drawn shut. I snapped the photo, and when I went back to look again at the photo, I got goosebumps"...unquote.

Genuine orbs are in constant motion, they have the ability to 'shape-shift' and can easily change their appearance from an opaque media to a translucent or transparent media.
This spiritual orb is quite large, in some areas it is semi-transparent, and in some areas it is completely transparent and almost invisible. It is hovering over this beautiful sleeping baby, and it has all the qualities and characteristics of a high-rank celestial Module 2 orb.

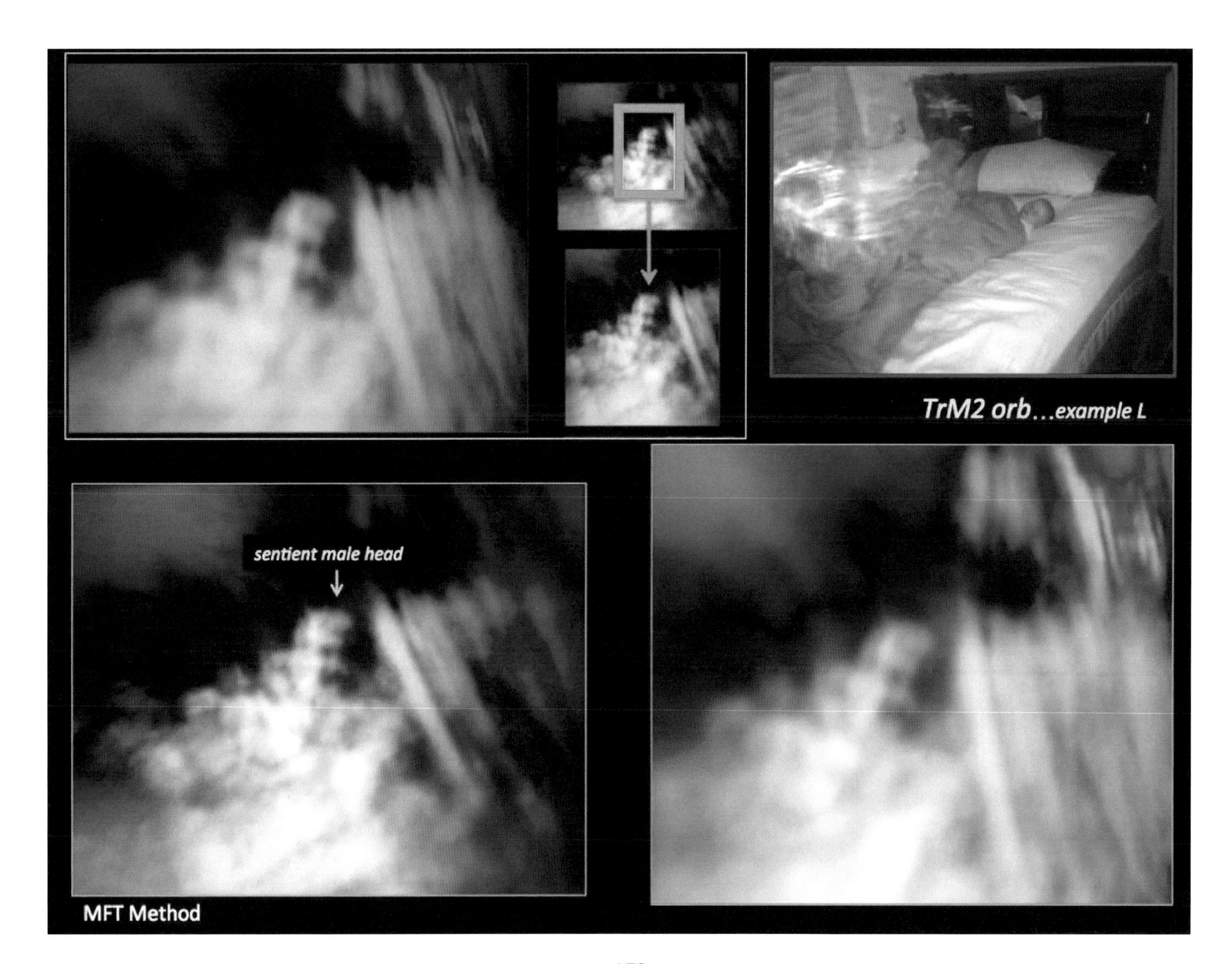
TrM2 orb…example L
sentient male head
MFT Method

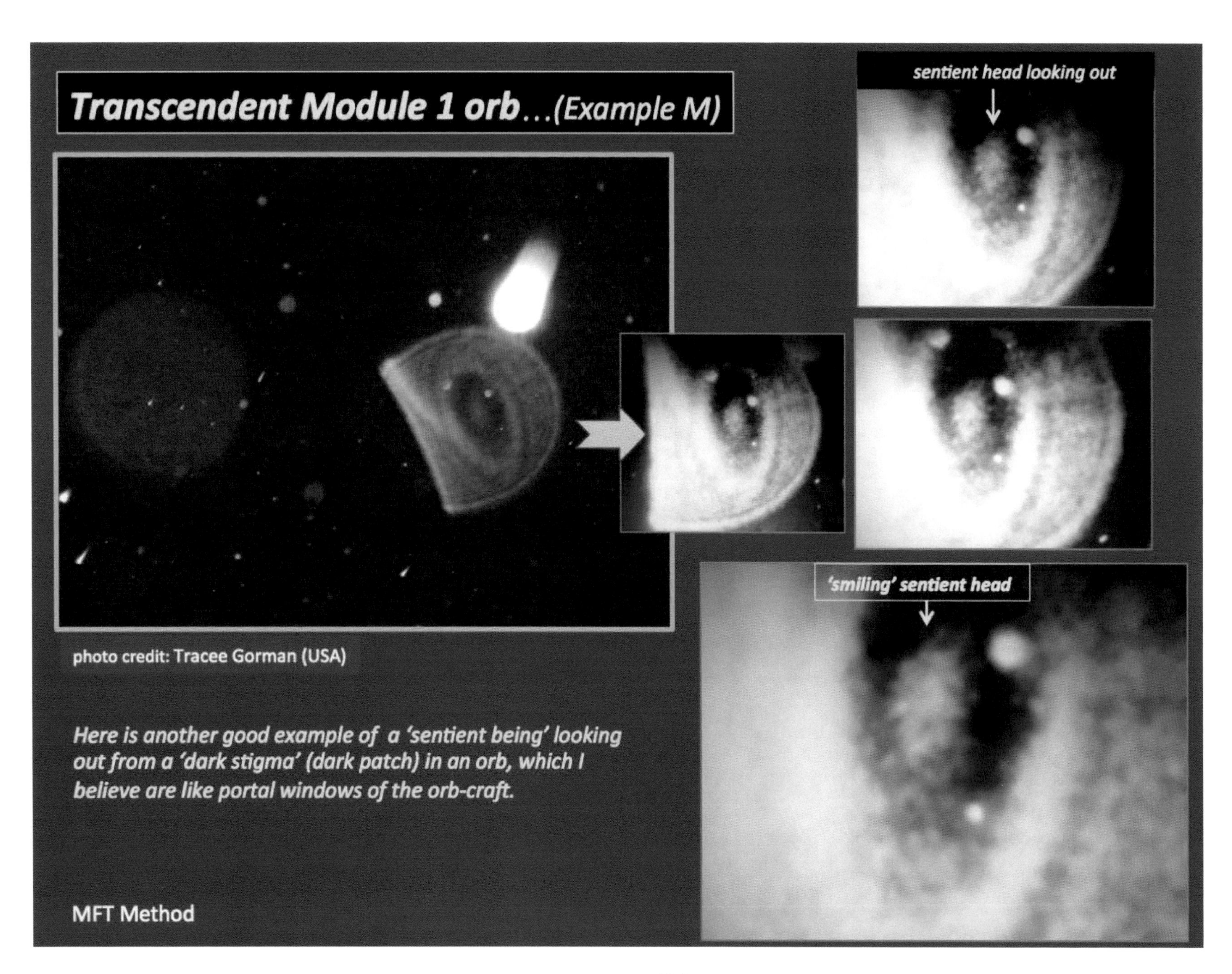
Transcendent Module 1 orb...(Example M)
sentient head looking out
'smiling' sentient head
photo credit: Tracee Gorman (USA)
Here is another good example of a 'sentient being' looking out from a 'dark stigma' (dark patch) in an orb, which I believe are like portal windows of the orb-craft.
MFT Method

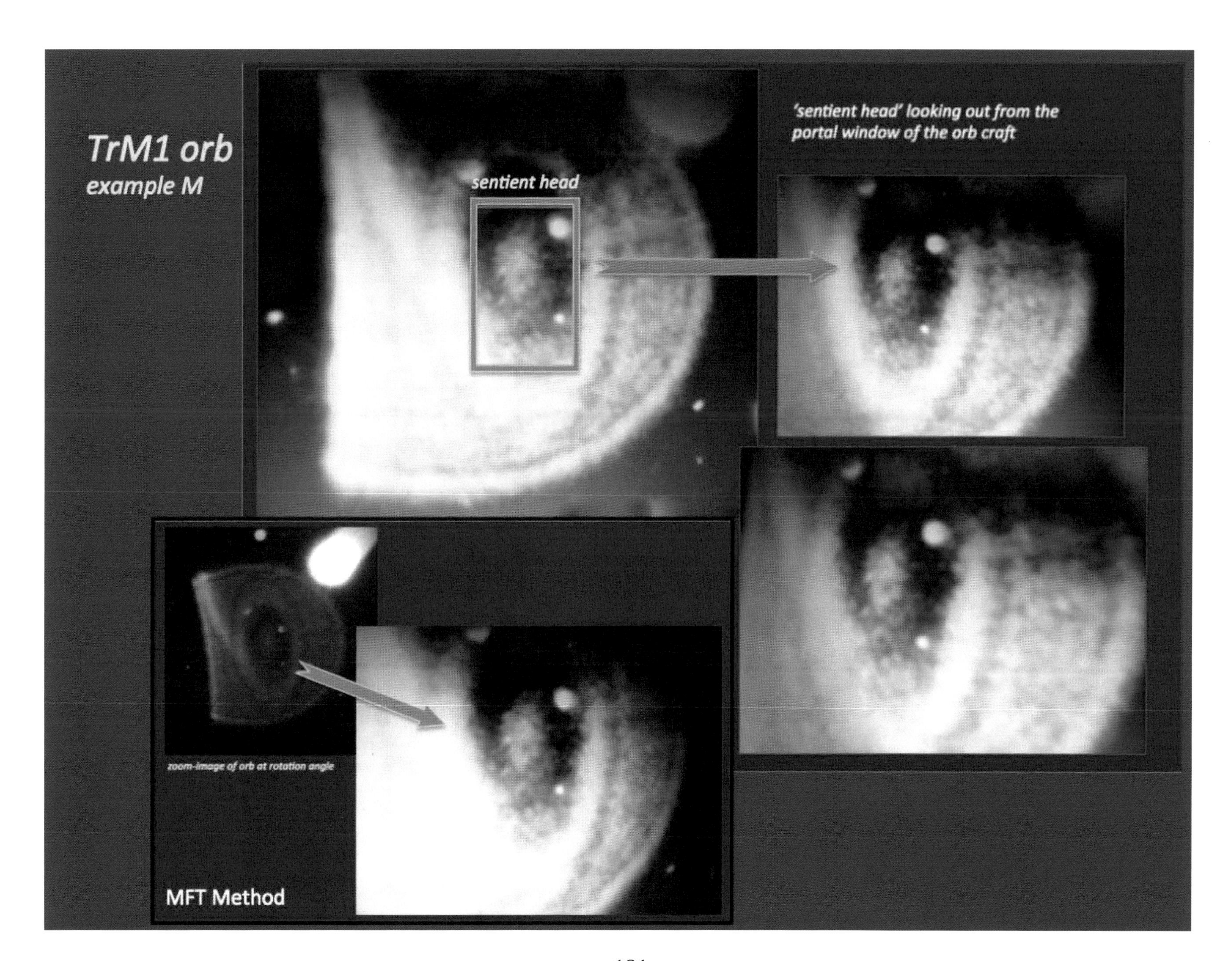
TrM1 orb
example M
sentient head
'sentient head' looking out from the portal window of the orb craft
zoom-image of orb at rotation angle
MFT Method

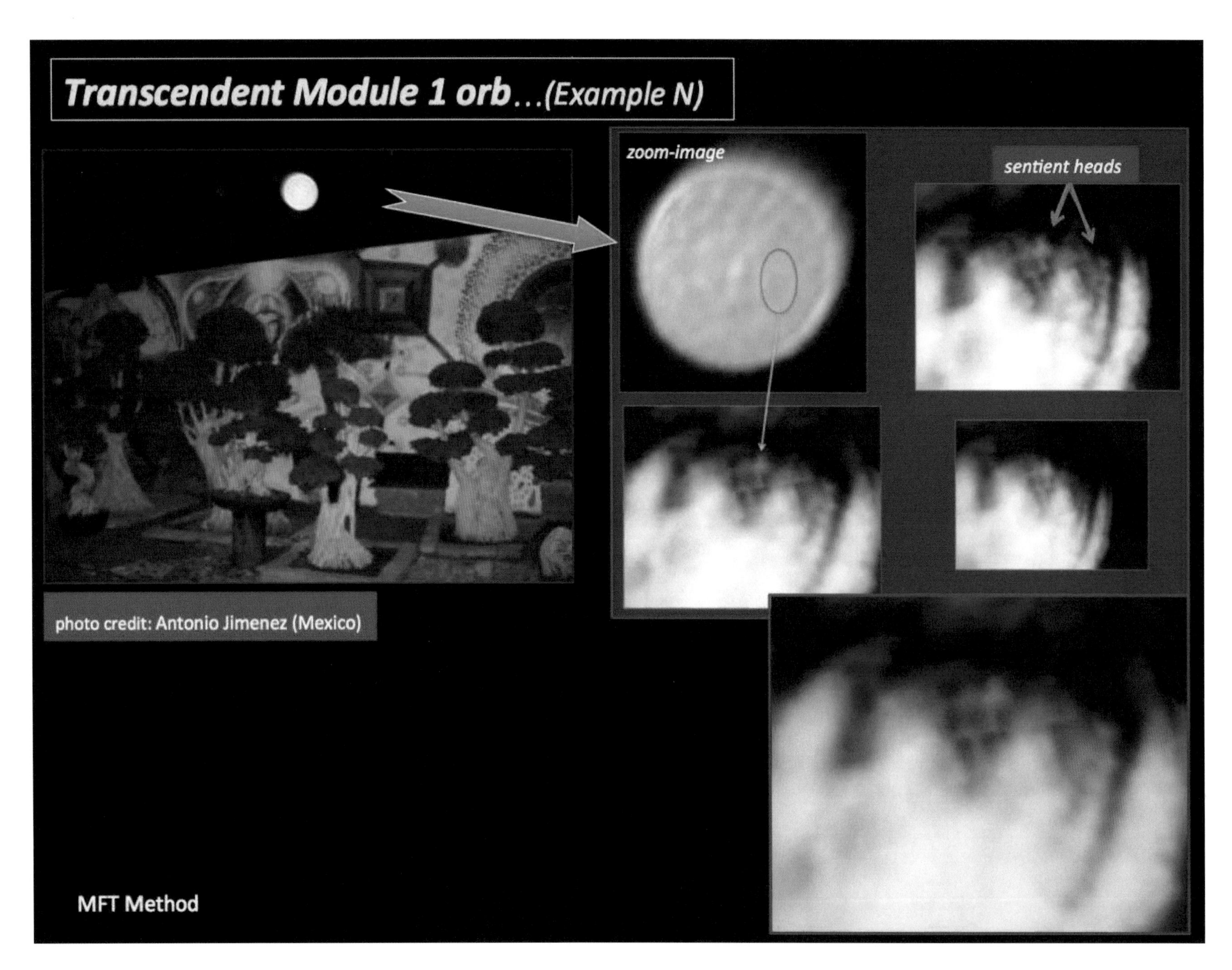
Transcendent Module 1 orb...(Example N)
zoom-image
sentient heads
photo credit: Antonio Jimenez (Mexico)
MFT Method

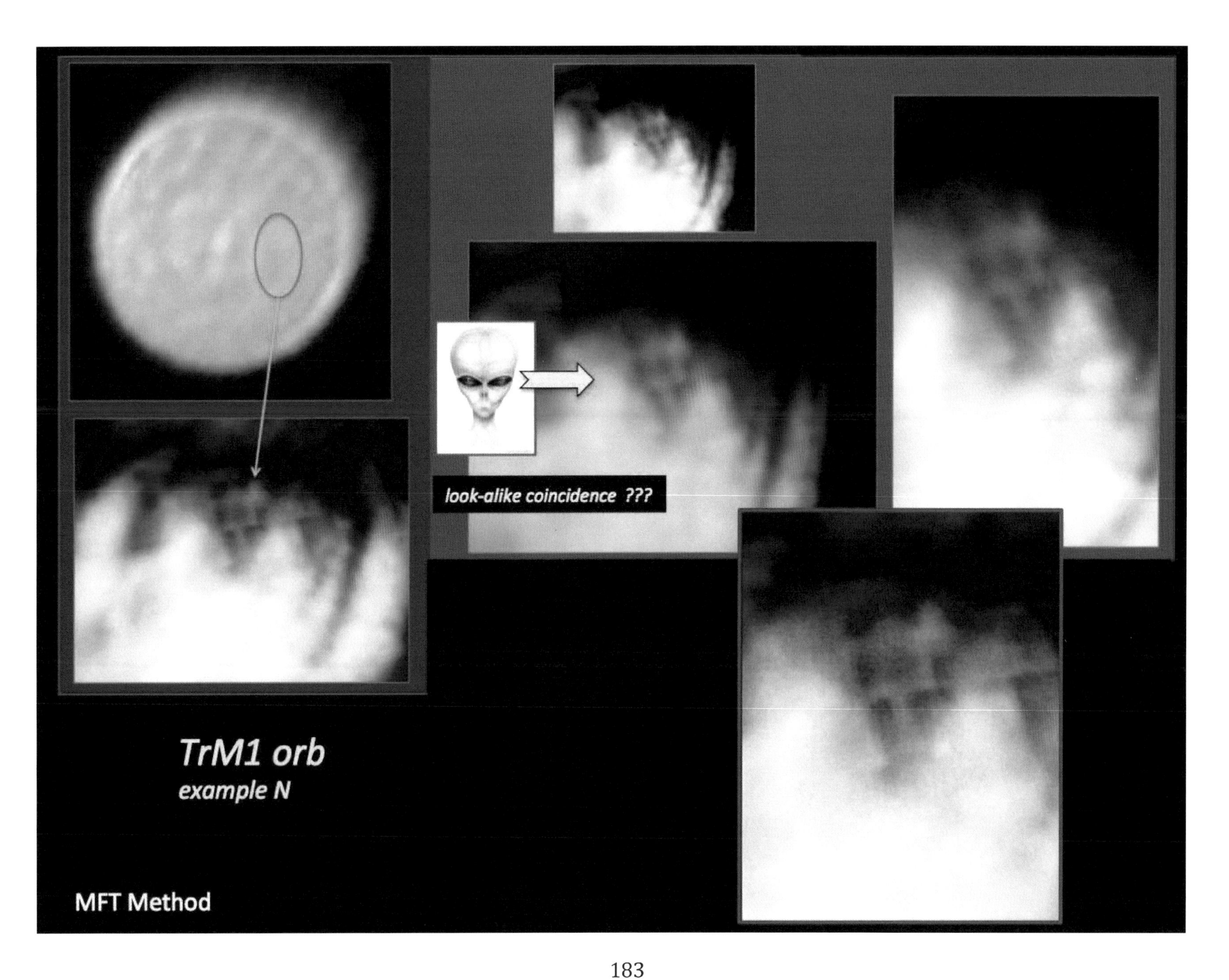
look-alike coincidence ???
TrM1 orb
example N
MFT Method

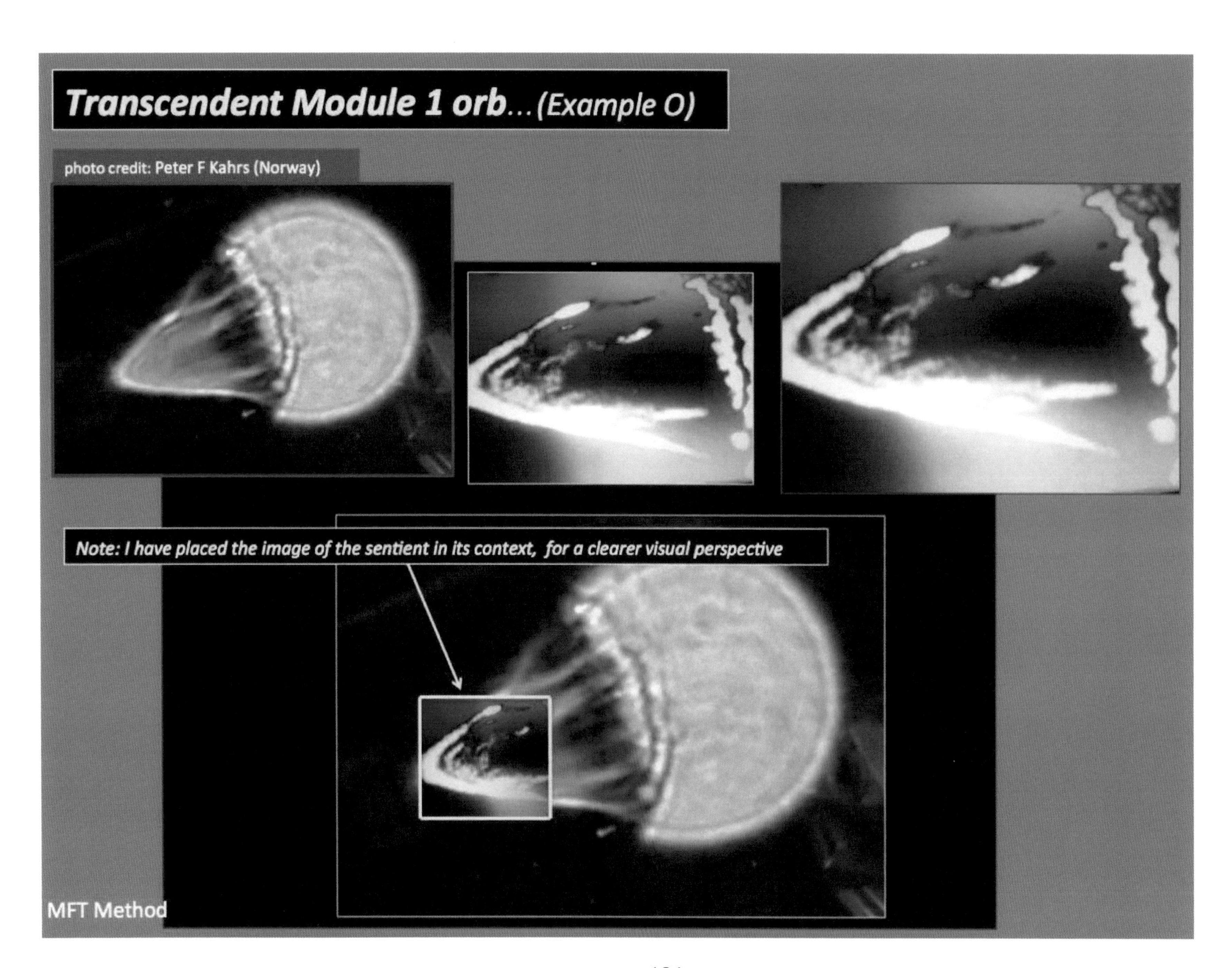
Transcendent Module 1 orb...(Example O)
photo credit: Peter F Kahrs (Norway)
Note: I have placed the image of the sentient in its context, for a clearer visual perspective
MFT Method

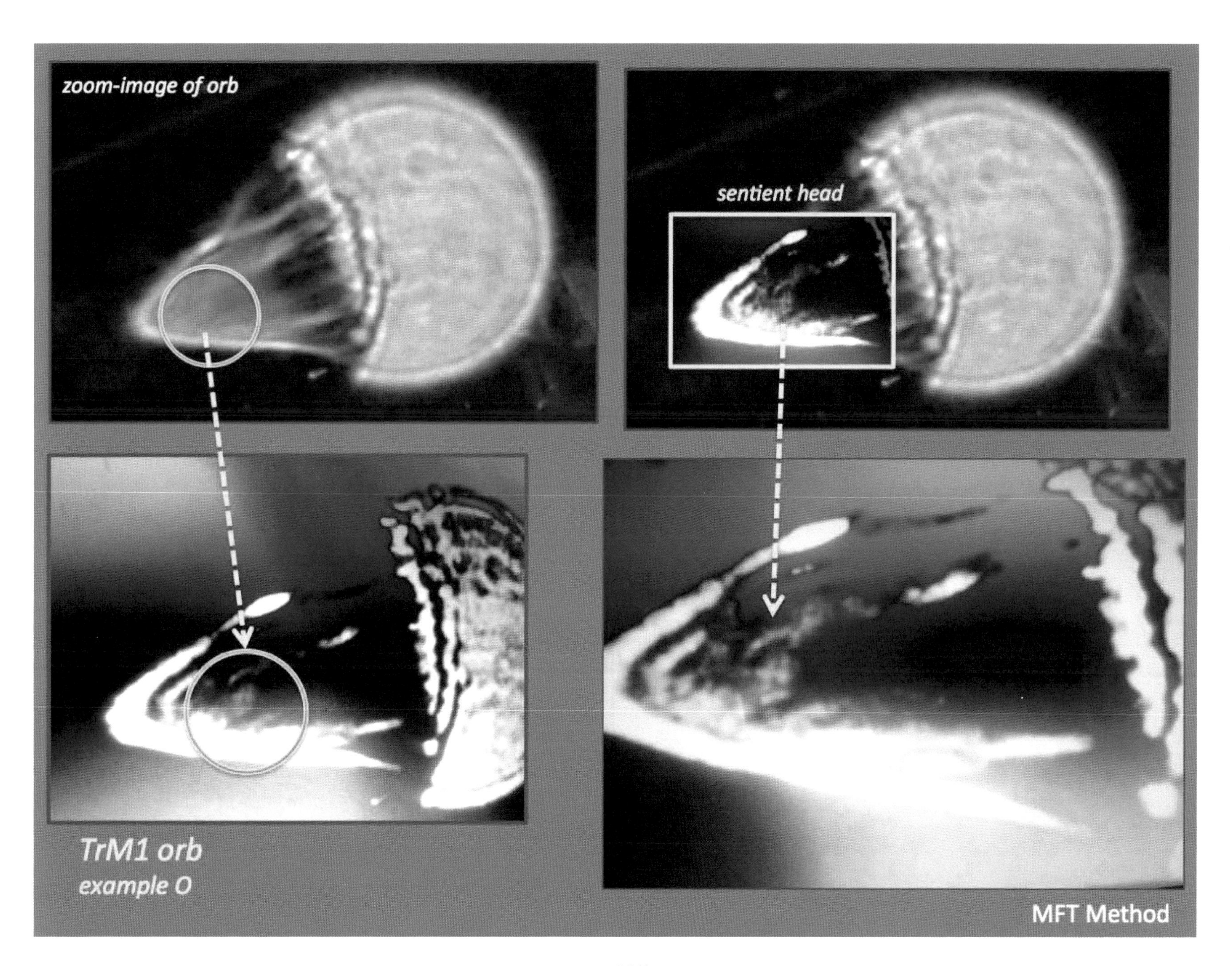
zoom-image of orb
sentient head
TrM1 orb
example O
MFT Method

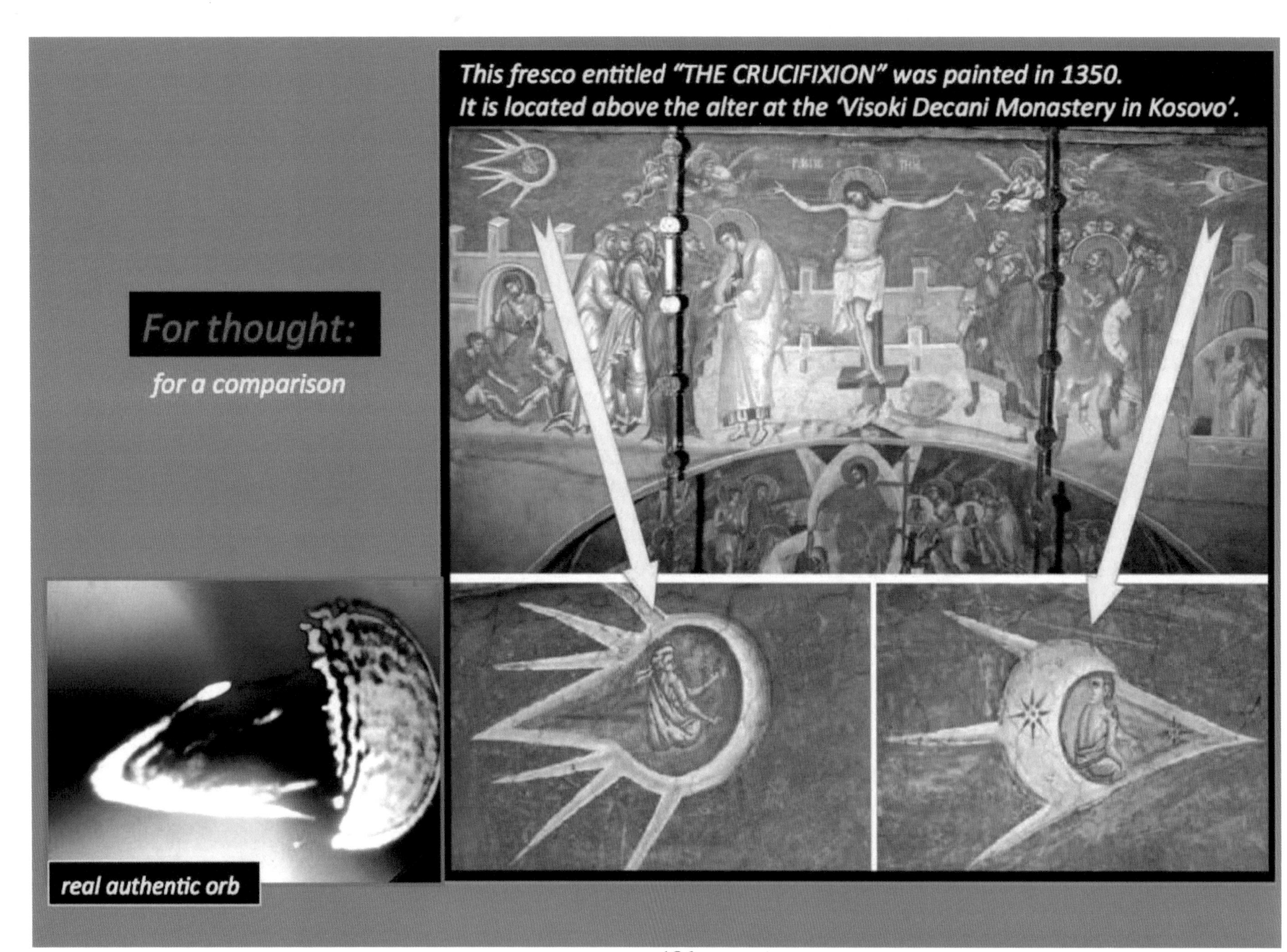
This fresco entitled "THE CRUCIFIXION" was painted in 1350.
It is located above the alter at the 'Visoki Decani Monastery in Kosovo'.
For thought:
for a comparison
real authentic orb

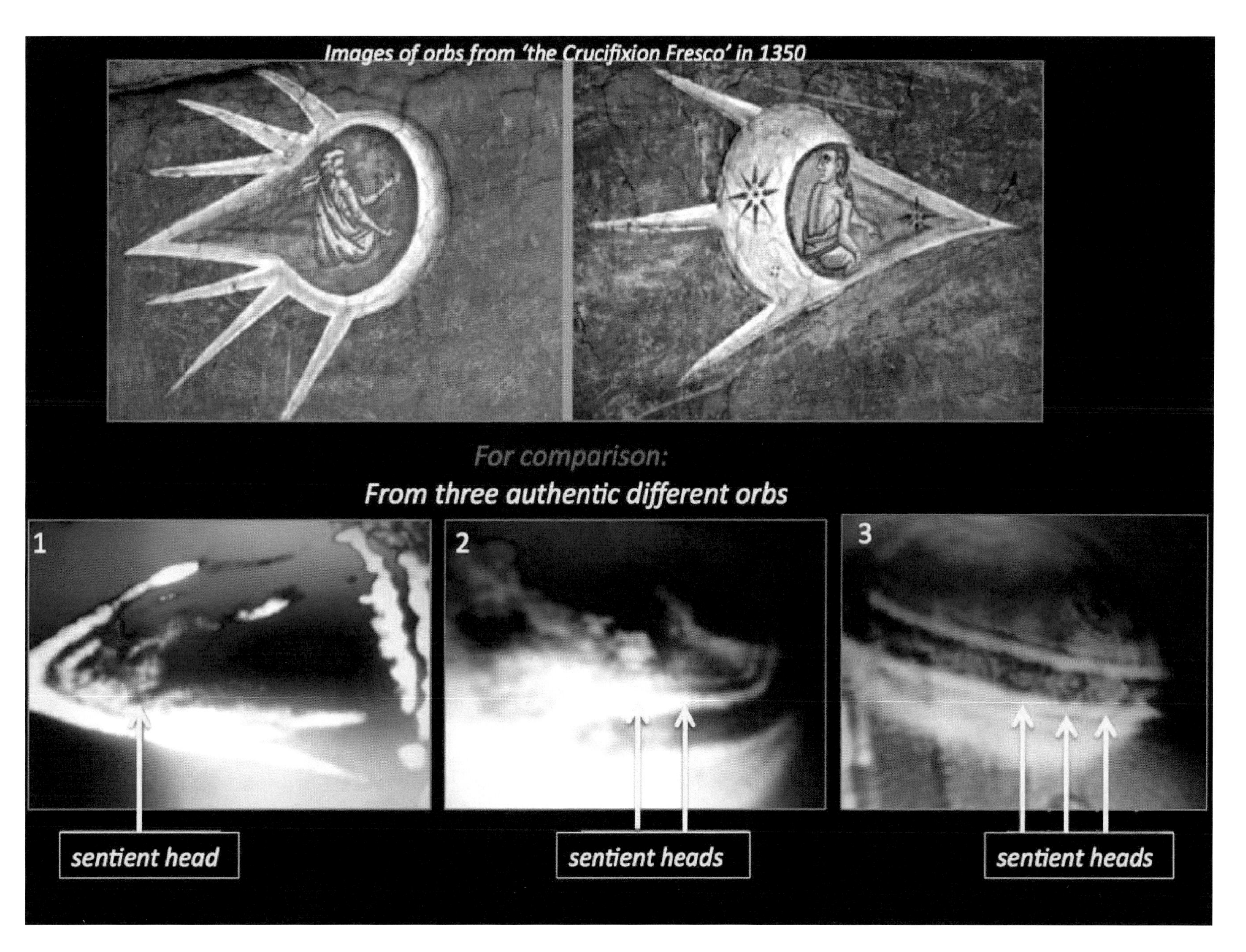
Images of orbs from 'the Crucifixion Fresco' in 1350
For comparison:
From three authentic different orbs
1
2
3
sentient head
sentient heads
sentient heads

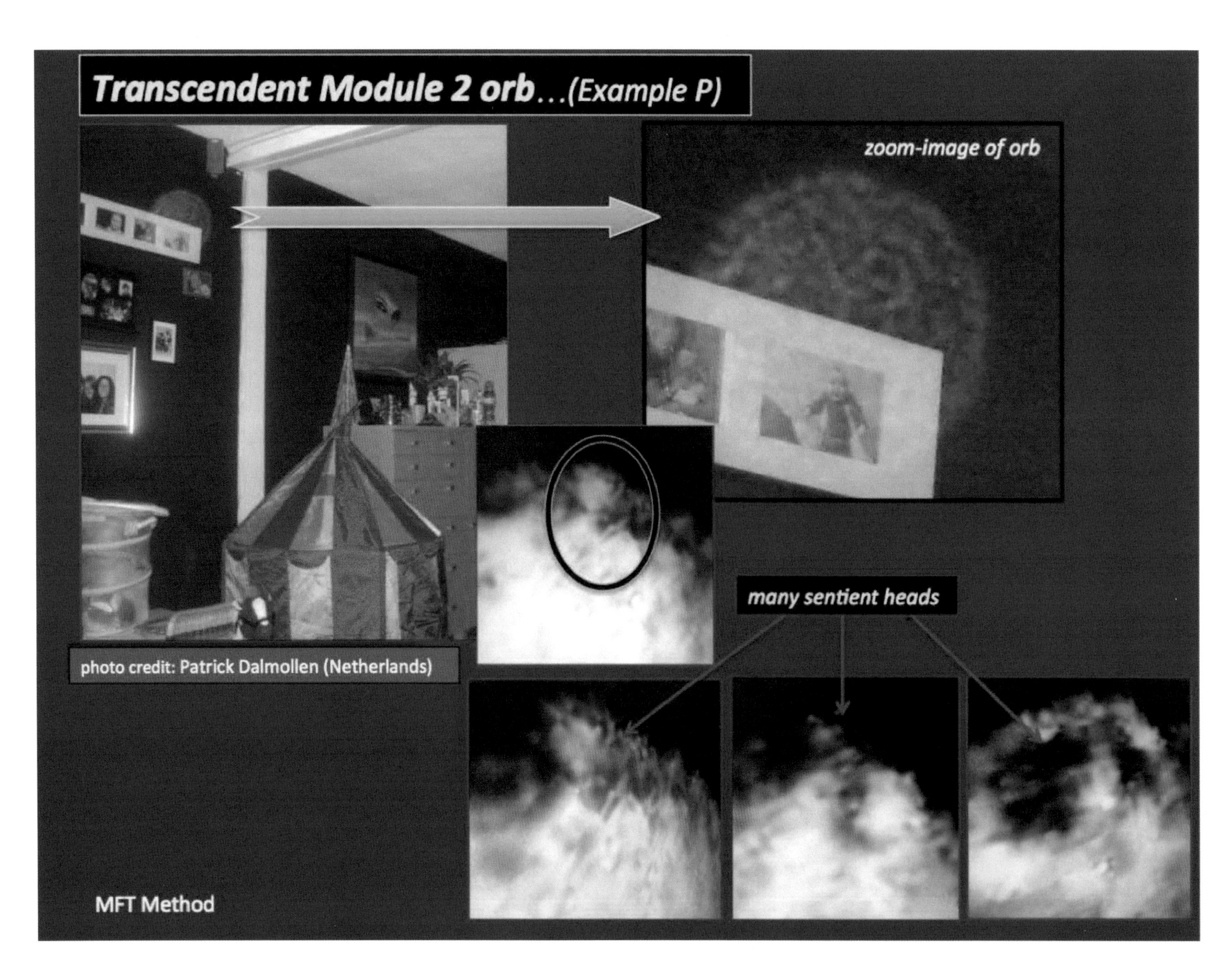
Transcendent Module 2 orb...(Example P)
zoom-image of orb
many sentient heads
photo credit: Patrick Dalmollen (Netherlands)
MFT Method

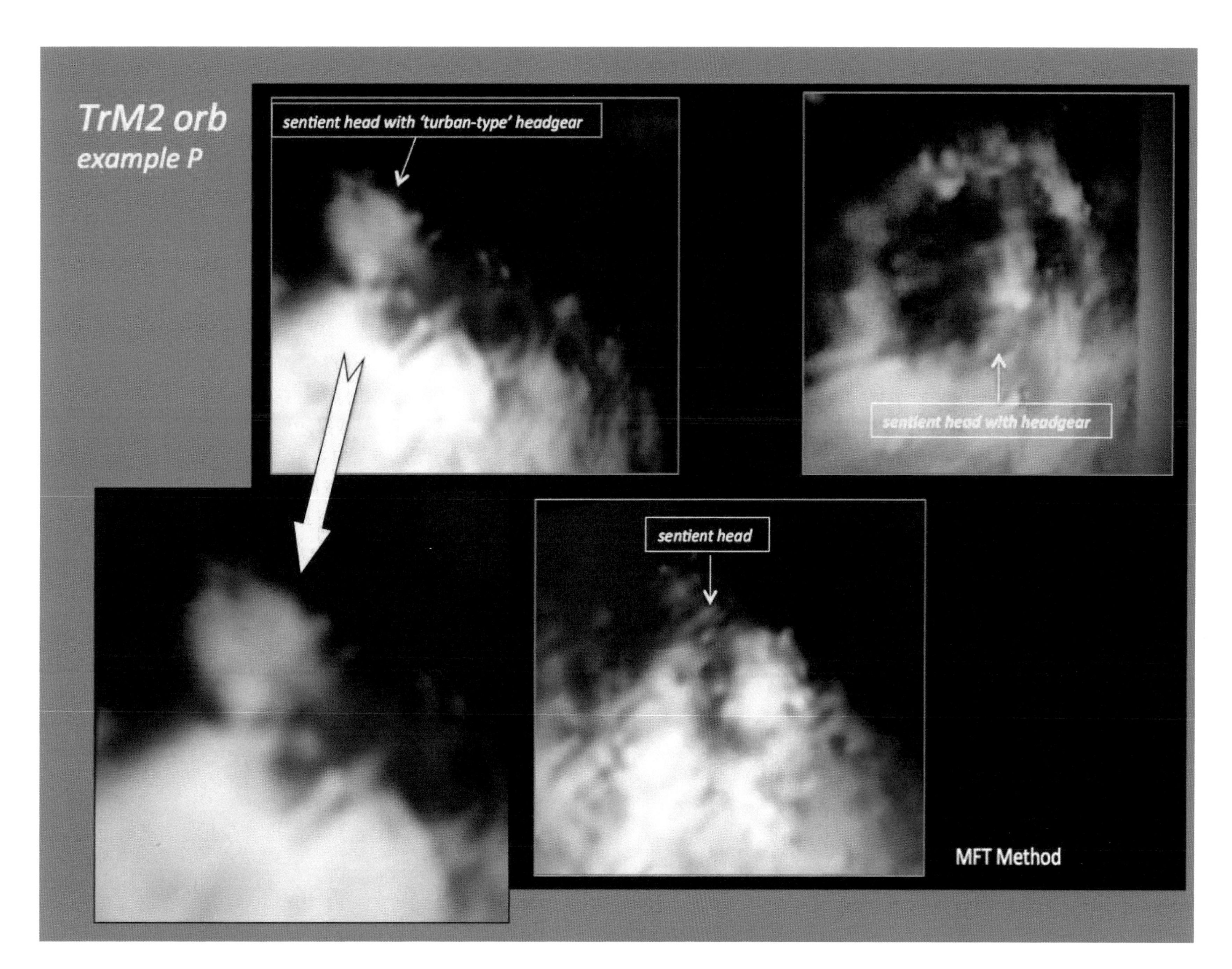
TrM2 orb
example P
sentient head with 'turban-type' headgear
sentient head with headgear
sentient head
MFT Method

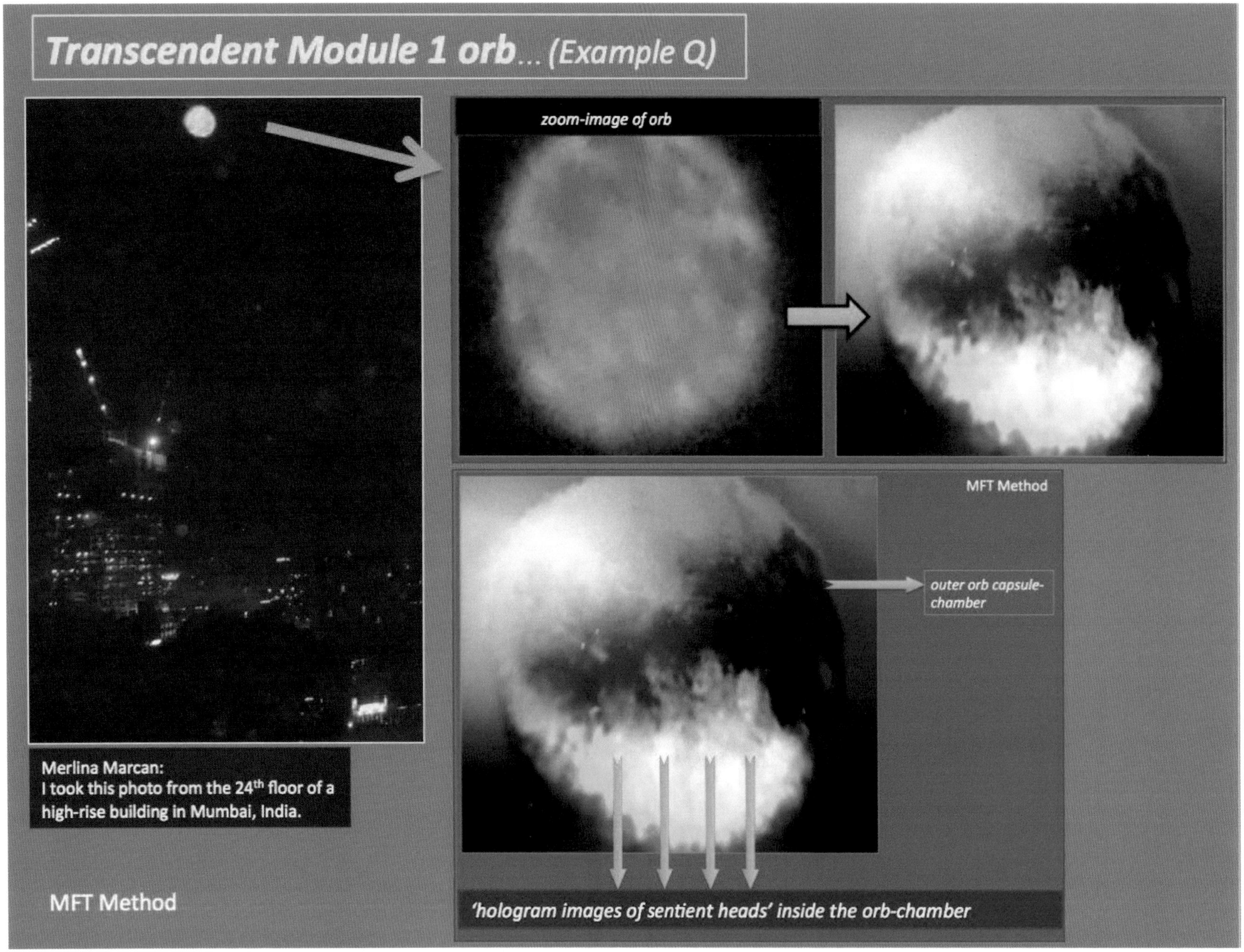

Transcendent Module 1 orb... (Example Q)

zoom-image of orb

MFT Method

outer orb capsule-chamber

'hologram images of sentient heads' inside the orb-chamber

Merlina Marcan:
I took this photo from the 24th floor of a high-rise building in Mumbai, India.

MFT Method

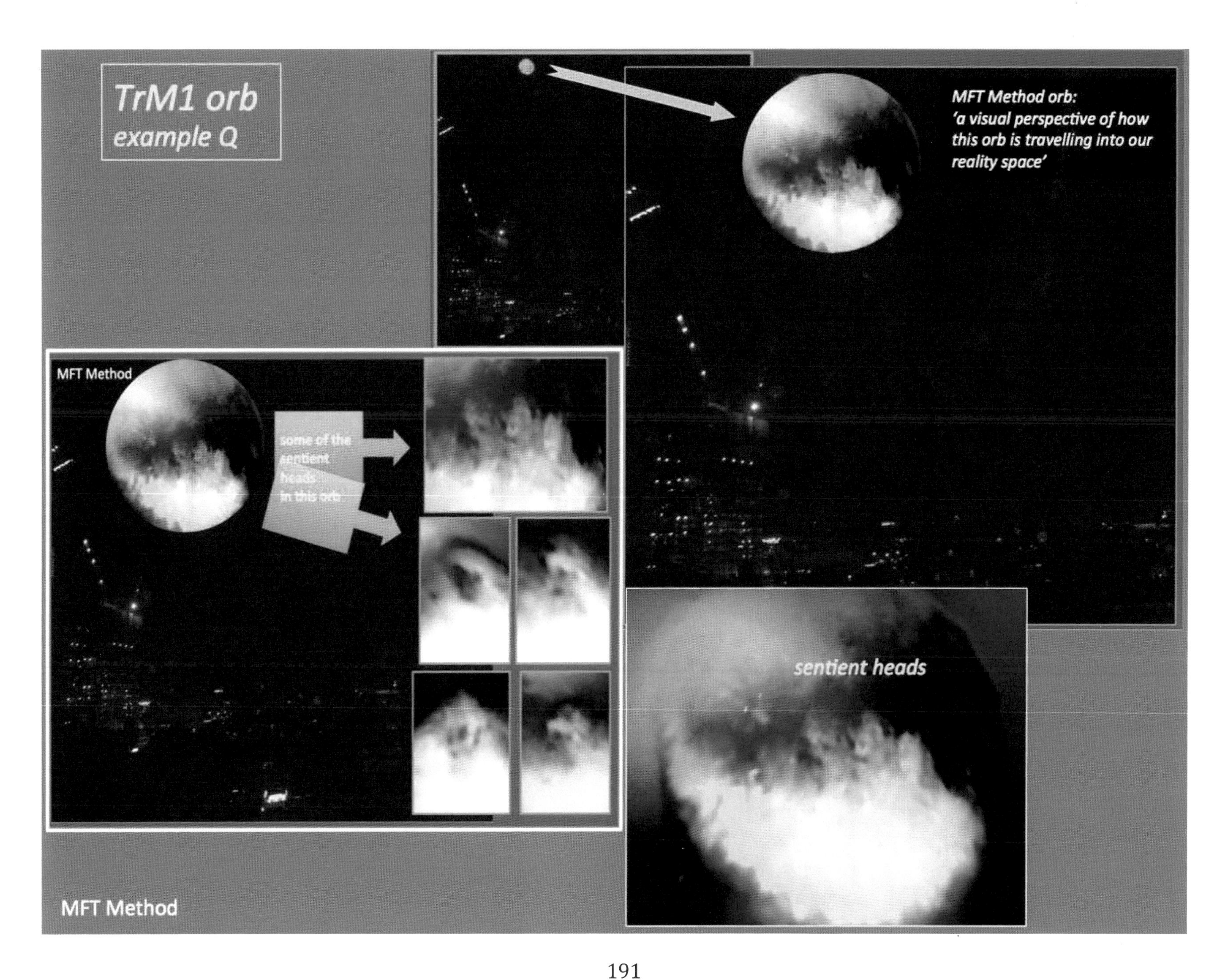
TrM1 orb
example Q
MFT Method orb:
'a visual perspective of how this orb is travelling into our reality space'
MFT Method
some of the sentient heads in this orb
sentient heads
MFT Method

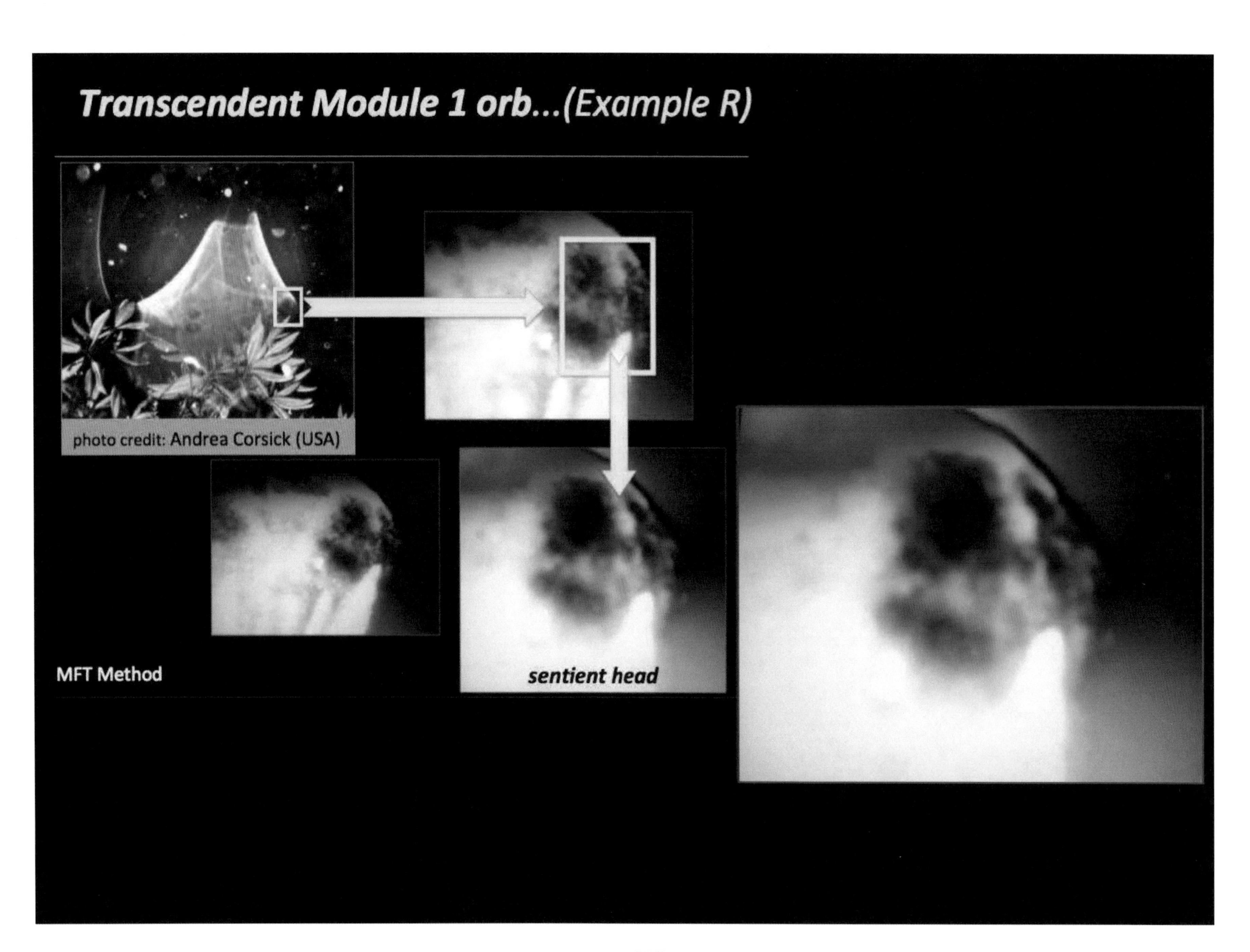
Transcendent Module 1 orb...(Example R)
photo credit: Andrea Corsick (USA)
MFT Method
sentient head

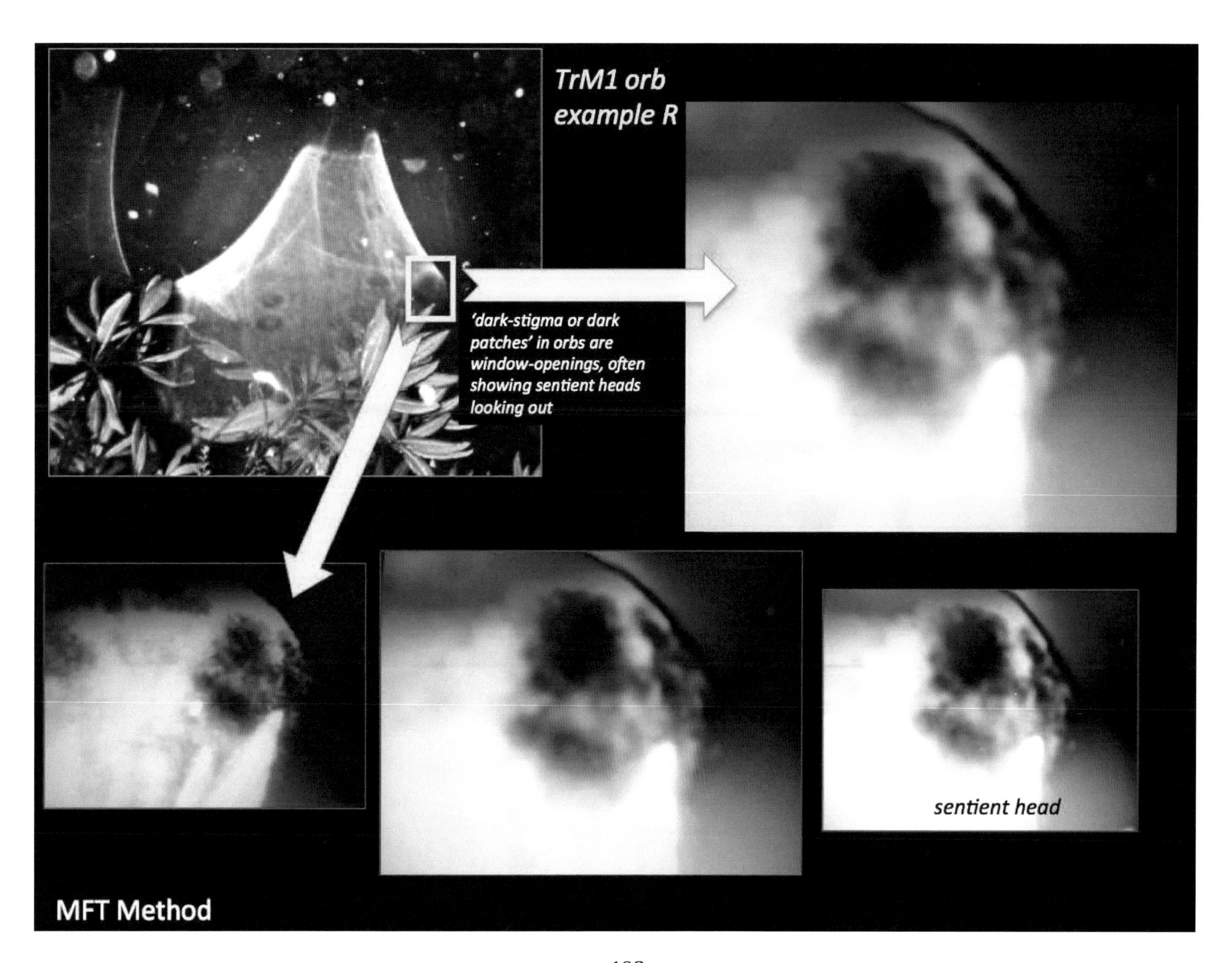
TrM1 orb
example R
'dark-stigma or dark patches' in orbs are window-openings, often showing sentient heads looking out
sentient head
MFT Method

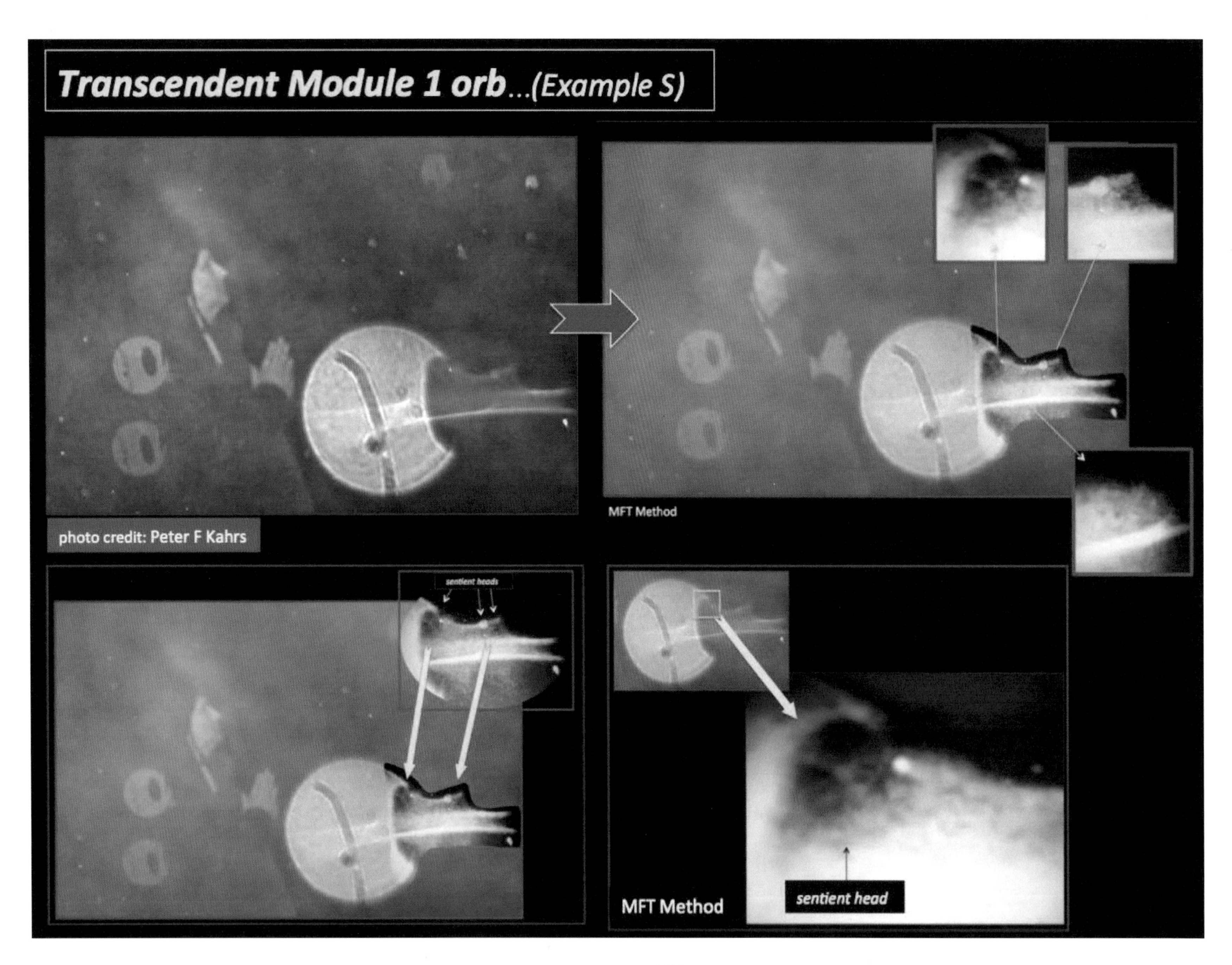
Transcendent Module 1 orb...(Example S)
photo credit: Peter F Kahrs
MFT Method
MFT Method
sentient head

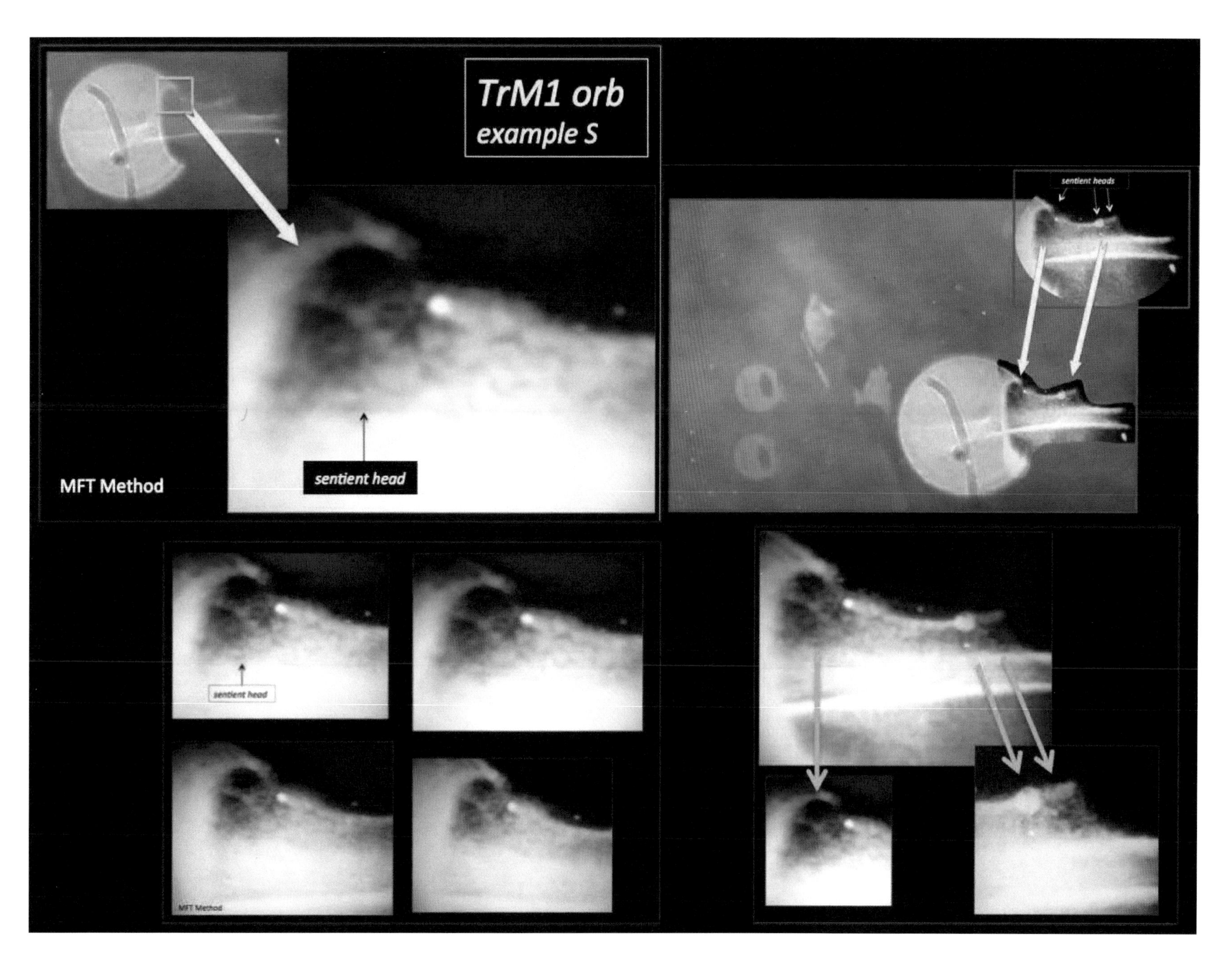
TrM1 orb
example S
sentient head
MFT Method
sentient heads
sentient head
MFT Method

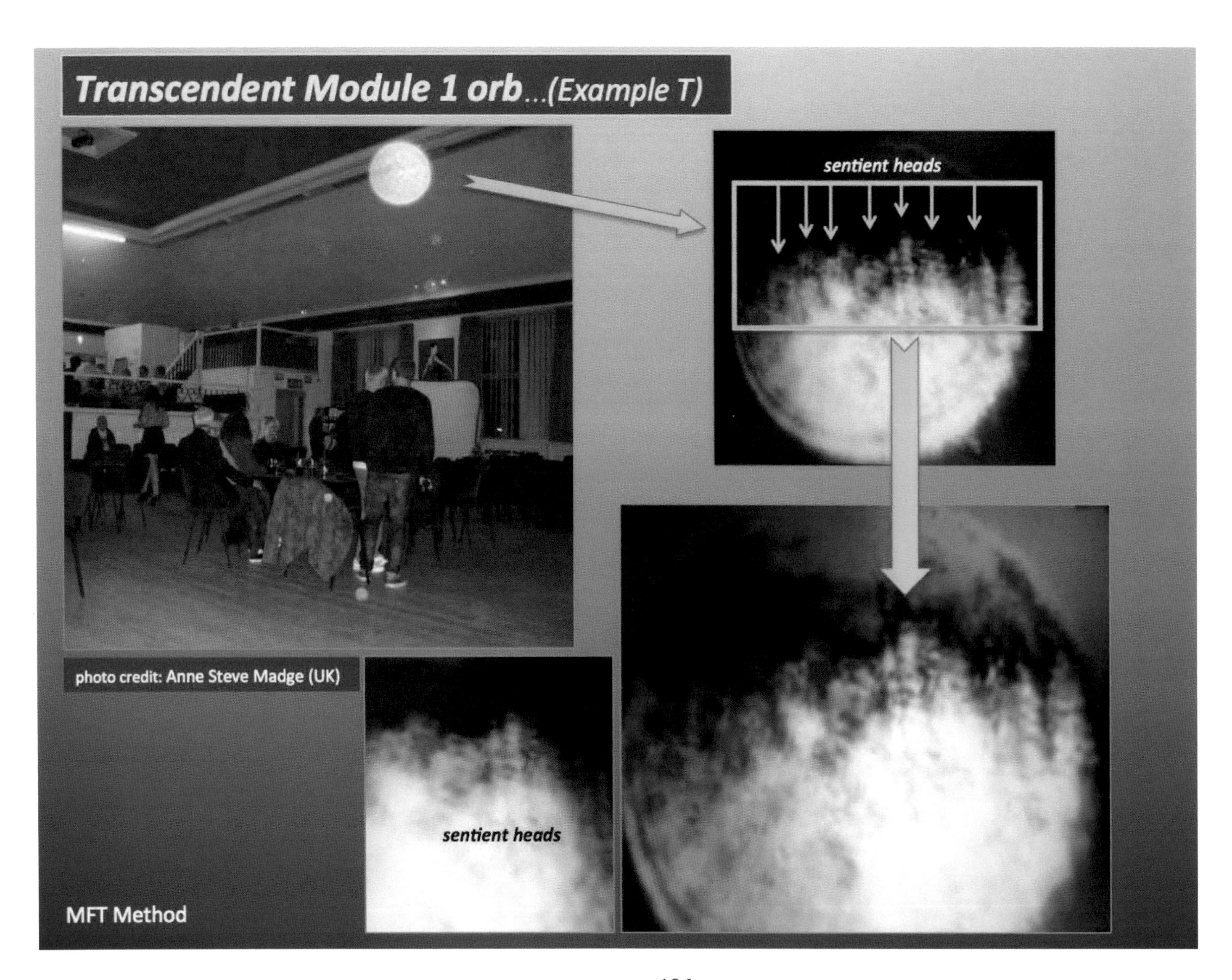
Transcendent Module 1 orb...(Example T)
sentient heads
photo credit: Anne Steve Madge (UK)
sentient heads
MFT Method

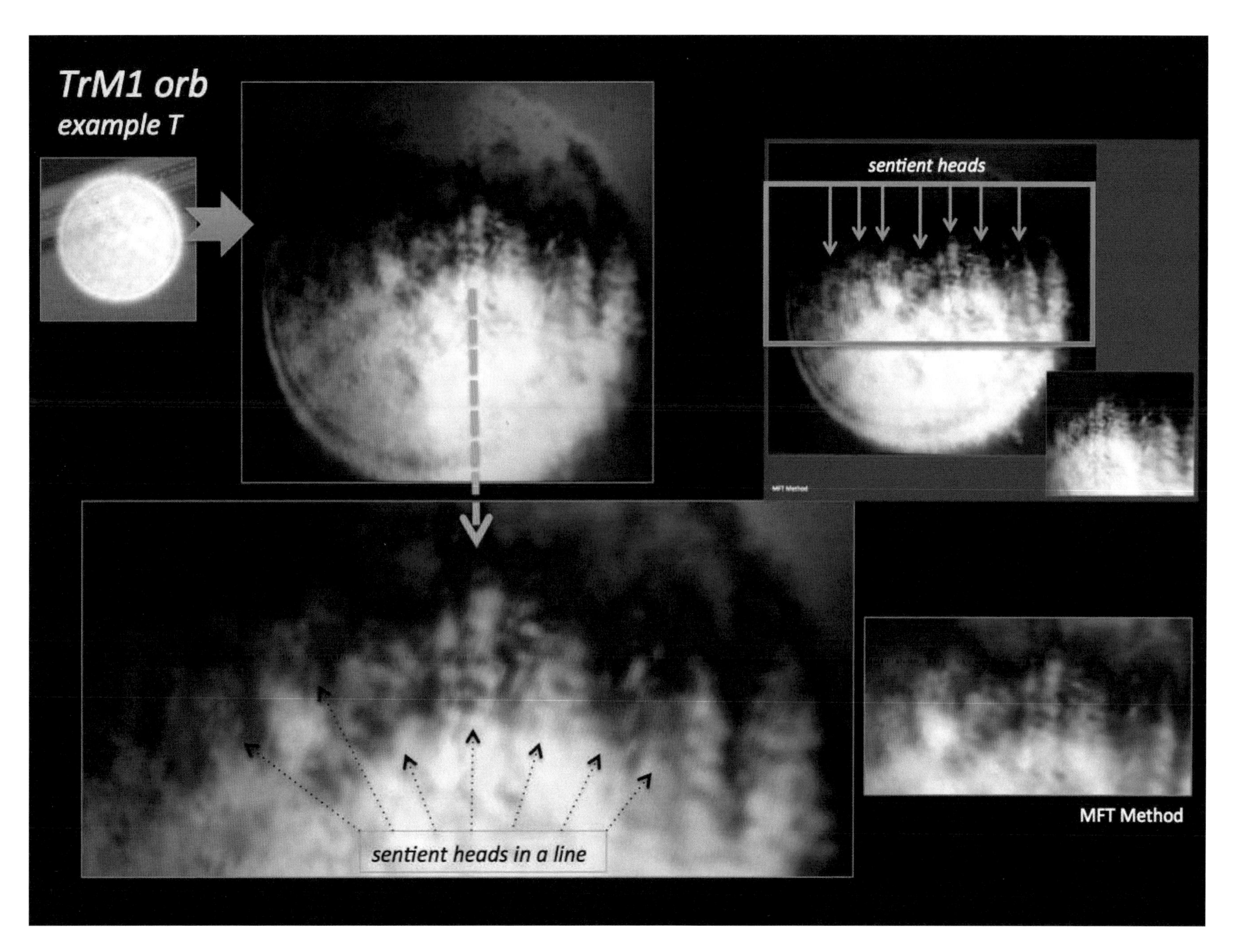
TrM1 orb
example T
sentient heads
sentient heads in a line
MFT Method

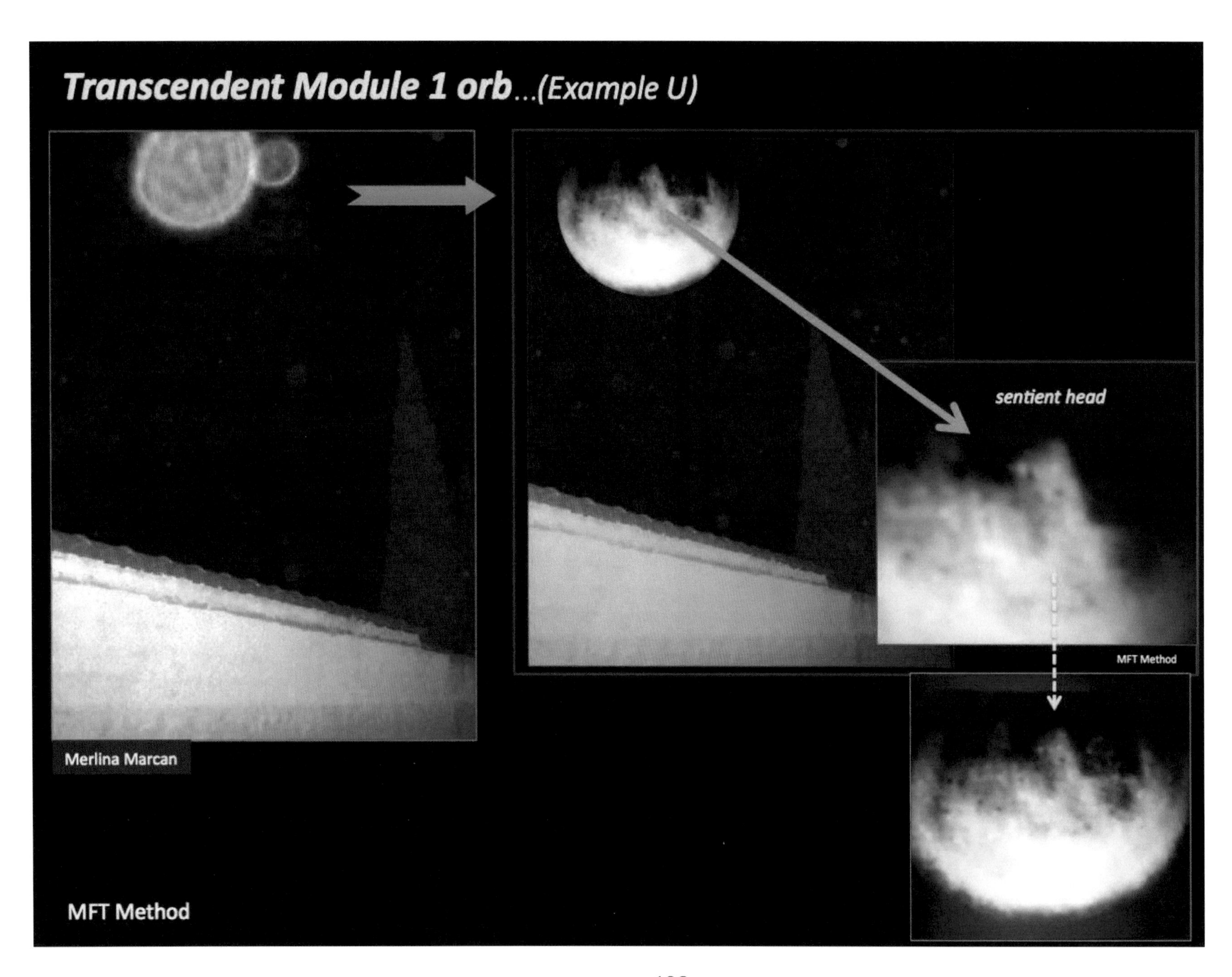
Transcendent Module 1 orb...(Example U)
sentient head
MFT Method
Merlina Marcan
MFT Method

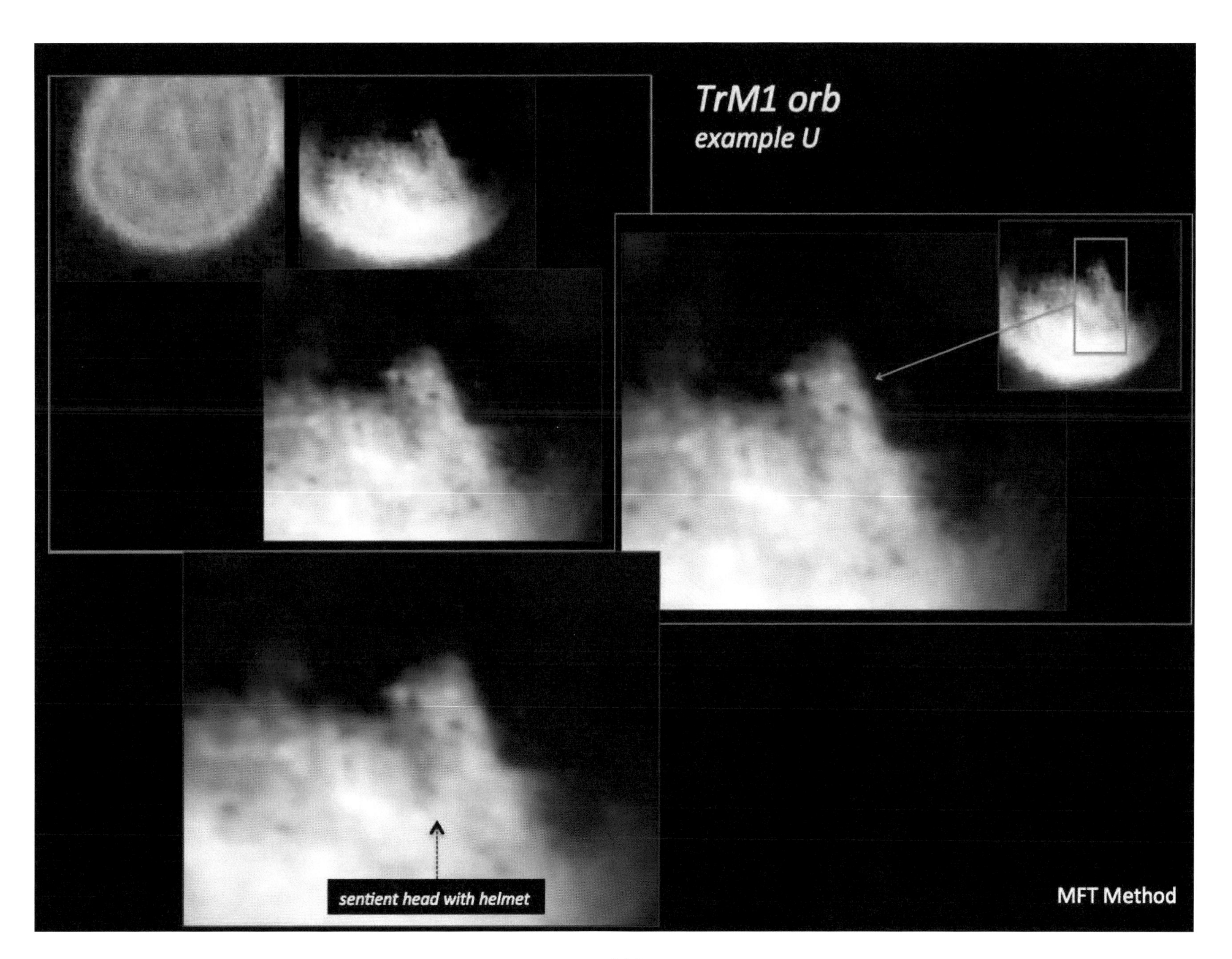
TrM1 orb
example U
sentient head with helmet
MFT Method

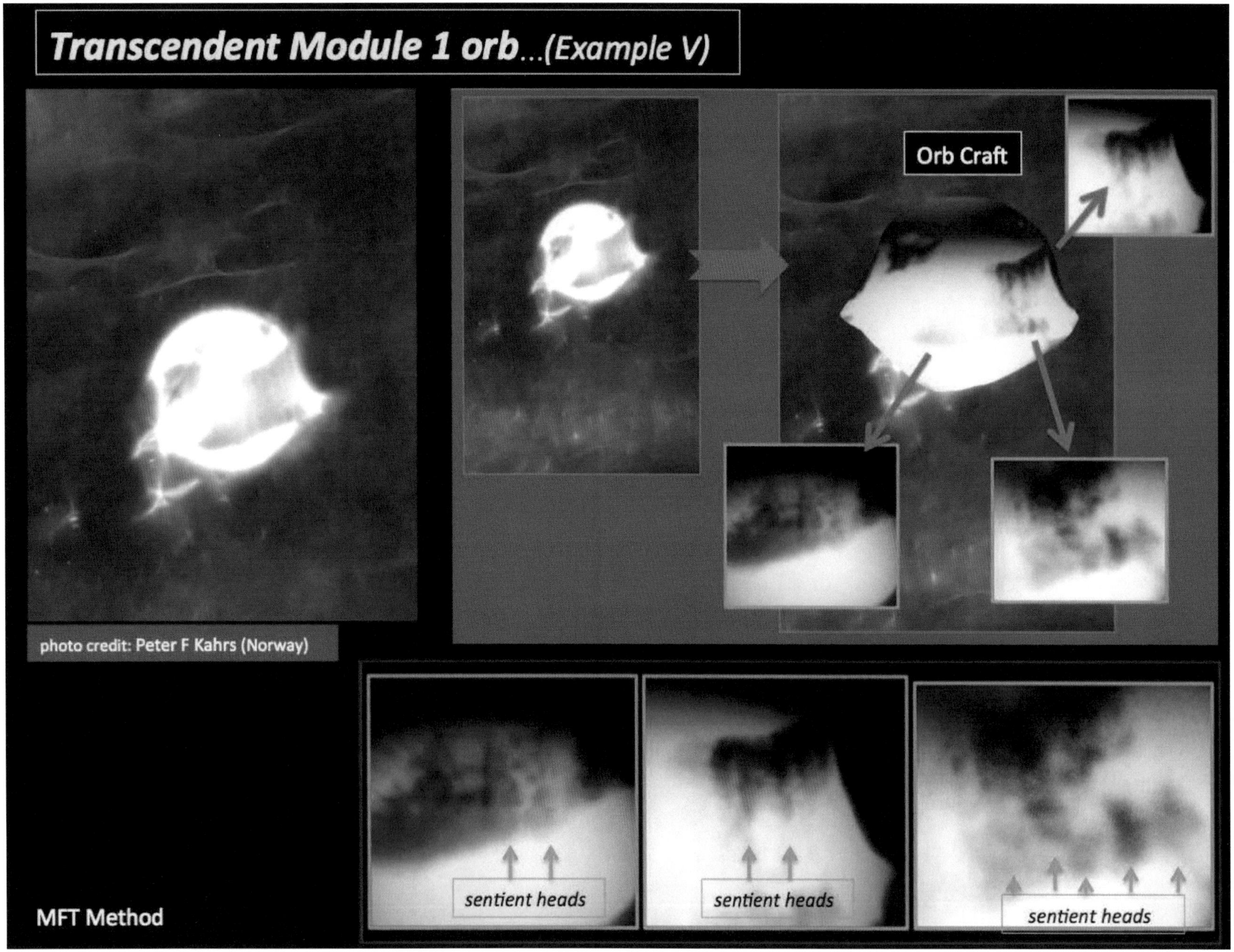

***Transcendent Module 1 orb**...(Example V)*

Orb Craft

photo credit: Peter F Kahrs (Norway)

sentient heads

sentient heads

sentient heads

MFT Method

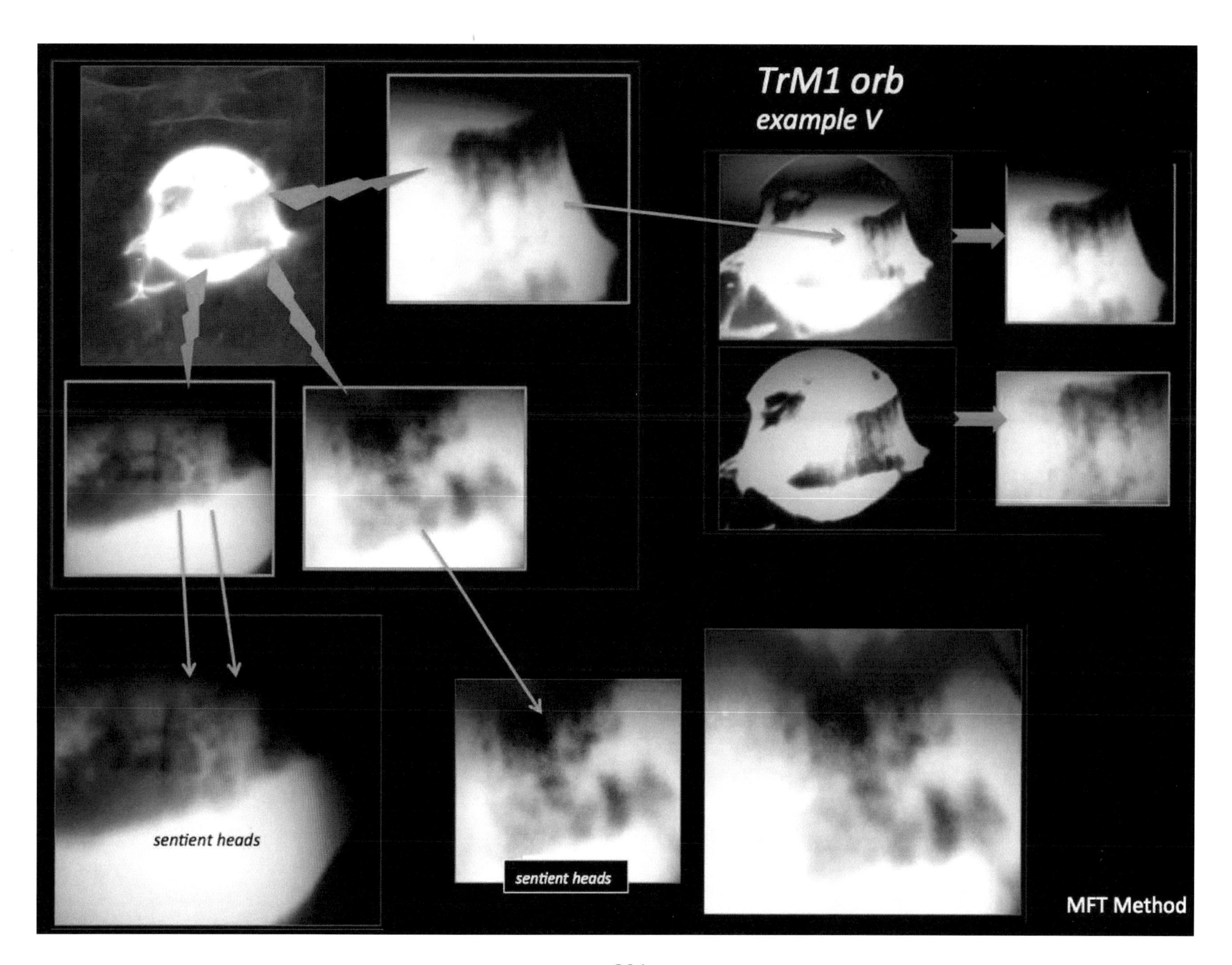
TrM1 orb
example V
sentient heads
sentient heads
MFT Method

Part 17
Real images of Sentient…'and my paintings'

I am not an artist, but I often find peace and relaxation when I sometimes have time to myself to paint.
I paint from my heart and enjoy giving away my artwork as little gifts to friends and family.

One day, a friend suggested that I should paint some of the *beings* that I was capturing with my 'MFT Method', and I thought that was a brilliant idea! So I did! I started to paint portraits of some of my most favourite beings in my collections, and here is the first series of *six* of these beings.
Needless to say, that some of my Research Work and Artwork is '*proudly hanging*' on my bedroom wall as a reminder to *never give up*! It is an on-going source of motivation for me every single day when I wake up and look at my wall, which is opposite my bed, that this is my *mission* and calling.

It is a stimulus that we *all* have a goal in this life, whether big or small and we need to strive to fulfill our dreams of whatever makes us happy, or gives us a positive fulfillment.
(See photos, PAGES 203-204)

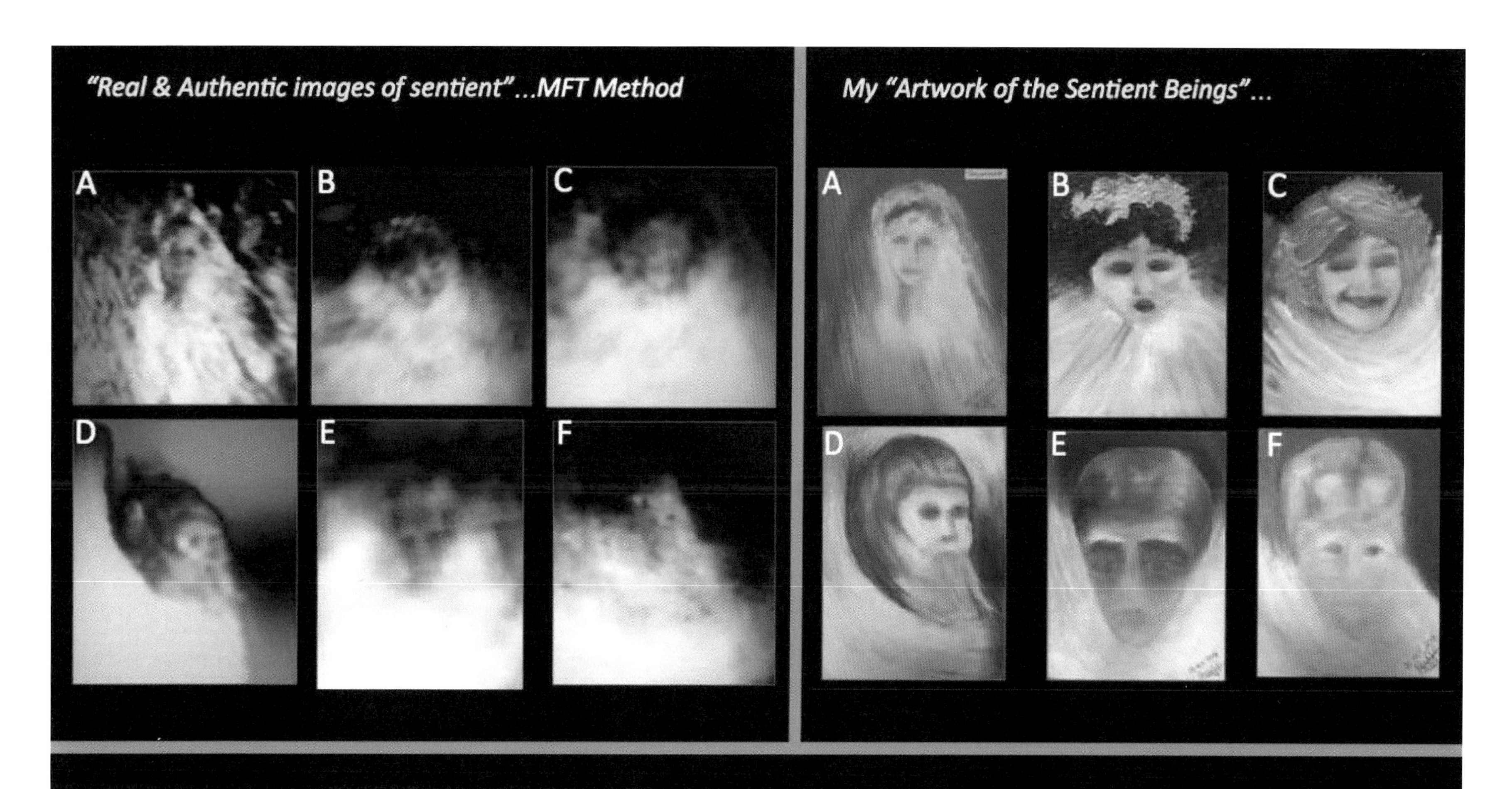

There are many species of 'Space Beings' that can easily glide and ride through our Earth's atmosphere...
These so called 'cosmic-travellers' or space beings are advanced civilizations, that can co-exist in, on, and above our planet Earth.

(I am not an artist, but I often find peace and relaxation when I sometimes have time to myself, to paint. I paint from my heart and my intuition, and the above images (on the right) are the final portraits of the Real Cosmic Travellers (on the left).

This is my 'Artwork' and my 'Research work' that is proudly hanging on my bedroom wall…

Part 18
Conclusion

As a conclusion to my book, I would like to say that '*we are being visited here on mother Earth by Beings from other dimensions, realities and other cosmic civilizations*. They travel in crafts that are capable of travelling distances of many Light Years away, they travel quickly and effectively, and their orb-crafts are some kind of energy or magnetic fields that can bypass our acceptable knowledge of physics!

Their mission is peaceful, they show signs of high intelligence, and they appear to want to communicate with us. There are multitudes of *different varieties of orb-crafts* that are coming from different areas of our universe, and from different civilizations of the cosmos. They are highly advanced civilizations and have highly advanced technologies; they have found a way to travel via what we may call ***'wormholes'***. Whether we call these powerful energy sources of travel; *wormholes, jump gates, black-holes, vortices or portals*, the important thing is that they have found a way to travel here to *us,* because they know we cannot travel to *them*.

It is a blessing to be able to SEE THEM, because they are allowing us to SEE THEM! They can choose 'when and where' they will allow us to SEE THEM. It is not surprising that they **do not** show up in photographs of professional photographers or professional films makers, because they very well know that they will not be welcome, therefore they do not show their presence. They seem to especially like to be photographed in loving, happy, calm, entertaining and enjoyable environments, and will often strategically position themselves near and around people, children, animals and nature.

Note:
I get confirmation from 'them' almost every day. I will finish this last page with a confirmation photo that I took at a 'UFO SKY WATCH' that we did at the STAR FAMILY CONFERENCE EVENT here in Australia.
We all gathered together after the conference on the beach to pray and meditate, and everyone sat forming a circle. While everyone began to 'chant Om', I quietly walked around the circle to take photos of our group, and '*I was asking them to show their presence*'.
This large *perfectly round orb* stopped its motion to a standstill, right in the middle of the group, only a few feet above our heads, to pose with us in the photo, confirming 'their presence', that they can see and hear us.
(This orb that appeared, is a TRANSCENDENT MODULE 2 orb, and I believe is from the *Spiritual Realms*).
(See photo, PAGE 206)

...THE END...

TrM2 orb

photo: Merlina Marcan

This photo was taken at a 'UFO Sky Watch at the STAR FAMILY CONFERENCE EVENT'.
Everyone sat forming a circle, to pray and meditate on the beach. So I quietly walked around the circle to take photos and I was asking them to show their presence. As soon as everyone 'chanted Om', this large orb appeared, right in the middle of the group, and only a few feet above our heads. (This beautiful large orb is perfectly round, which means that it stopped its motion to a stand-still, right in the middle of the group to pose for us in the photo, so as to confirm 'their presence' that 'they can see and hear us'.

Photo credits in this book:

A special thanks to the people listed below from all around the world, who have photographed or have been photographed with orbs around them, as the orbs 'wanted to appear in their photos'.

(List is in '*first-name*' alphabetical order):

Andrea Corsick (USA)
Anne Steve Madge (UK)
Antonino Lo Grasso (Italy)
Antonio Jimenez (Mexico)
Babie Thi Hatfield (USA)
Brittmarie Johansen (Sweden)
Cathy Finch (UK)
Cynthia Rae (USA)
Dawn Blackburn (UK)
Diana Davatgar (Germany)
Engel, die Geistige Welt (Germany)
Gila Carnaco (Italy)
Graca Pimentel (USA)
Isac Hdz
Jacqueline White (UK)
James B Lindsay (USA)
Sunshine Lanning (USA)
Juan Ishtar Abdiel (Mexico)
Karen Wilken (USA)
Kat J Peck (UK)
Kathy Jeffries
Liza Mdc (Australia)
Lama Serpo (Bhutan)
Marion Atehsa Cyrus (Germany)
Maratt Sandoval (Mexico)
Maya Bordjoski (Serbia)
Melita Jelenic
Michelle P (Australia)
Patrick Dalmollen (Netherlands)
Paul Greco (USA)
Peter F Kahrs (Norway)
Rickie Tan (Malaysia)
Rockette Marie (USA)
Ronnie R de Wit (Netherlands)
Tanya Trenouth (Canada)
Theresa Kaplan Amuso (USA)
Tracee McGrath Gorman (USA)

CPSIA information can be obtained at www.ICGtesting.com
Printed in the USA
BVIW121538210620
581956BV00005B/253